A BUILDING INFORMATION MODELLING MATURITY MODEL FOR DEVELOPING COUNTRIES

This book provides a reference point for the development of Building Information Modelling (BIM) maturity in the developing country context. Developing countries have been observed to have low BIM maturity and are struggling to adopt the technology amidst no clearly defined pathways for achieving BIM capability maturity. The research presented in this book provides construction industry stakeholders in developing countries with a framework and nomological map to aid in the advancement of BIM implementation. This work provides a pathway for overcoming the challenges inhibiting BIM maturity in developing countries and ultimately its diffusion in order to harness the benefits. The authors provide critical theoretical insights on BIM maturity in the developing country context, a comparative analysis of BIM maturity in both developing and developed countries, and finally, a conceptualisation of BIM maturity for developing countries. The book is unique as its construct is rooted in the state-of-the-art information management standards in the digitalisation era in the construction industry (ISO 19650). This book delivers a theoretical reference point to the academic and research community, and for the industry stakeholder, it acts as an essential guide to achieving BIM maturity at macro and micro levels.

Samuel A. Adekunle is a professional quantity surveyor and project manager and a post-doctoral research fellow at the Sustainable Human Settlement and Construction Research Centre, University of Johannesburg, South Africa.

Clinton Ohis Aigbavboa is a professor in the Department of Construction Management and Quantity Surveying and Director of cidb Centre of Excellence & Sustainable Human Settlement and Construction Research Centre, University of Johannesburg, South Africa.

Obuks Ejohwomu is Senior Lecturer in Project Management in the Department of Mechanical, Aerospace and Civil Engineering at the University of Manchester, United Kingdom.

Wellington Didibhuku Thwala is Director, School of Engineering and Distinguished Research Professor in the College of Science, Engineering and Technology, University of South Africa (UNISA), South Africa. Professor Thwala has extensive experience in providing consultancy for project leadership and management of construction engineering projects and teaching project management subjects at postgraduate level. He has varied research interests, including project management, construction engineering, supply chain management, leadership in construction engineering, Industry 4.0 and Smart Cities. Professor Thwala is the Editor-in-Chief of the *Journal of Construction Project Management and Innovation* (JCPMI) and serves as an editorial board member to various reputable international journals. He is a National Research Foundation (NRF) Rated Researcher.

Mahamadu Abdul-Majeed is Associate Professor of Innovative and Industrialised Construction at The Bartlett School of Sustainable Construction UCL, United Kingdom.

Routledge Research Collections for Construction in Developing Countries
Series Editors: Clinton Aigbavboa, Wellington Thwala, Chimay Anumba, David Edwards

A Building Information Modelling Maturity Model for Developing Countries

Samuel A. Adekunle, Clinton Aigbavboa, Obuks Ejohwomu, Wellington D. Thwala, and Mahamadu Abdul-Majeed

Routledge
Taylor & Francis Group

LONDON AND NEW YORK

First published 2023
by Routledge
4 Park Square, Milton Park, Abingdon, Oxon OX14 4RN

and by Routledge
605 Third Avenue, New York, NY 10158

Routledge is an imprint of the Taylor & Francis Group, an informa business

British Library Cataloguing-in-Publication Data
A catalogue record for this book is available from the British Library

Library of Congress Cataloging-in-Publication Data
Names: Adekunle, Samuel A., author. | Aigbavboa, Clinton, author. | Ejohwomu, Obuks, author. | Thwala, Wellington, author. | Abdul-Majeed, Mahamadu, author.
Title: A building information modelling maturity model for developing countries / Samuel A. Adekunle, Clinton Aigbavboa, Obuks Ejohwomu, Wellington D. Thwala, Mahamadu Abdul-Majeed.
Description: Abingdon, Oxon; New York, NY: Routledge, 2023. | Series: Routledge research collections for construction in developing countries | Includes bibliographical references and index. | Summary: "This book adds important literature to the BIM maturity body of knowledge in developing countries and it fills huge gap in research and industry practice"— Provided by publisher.
Identifiers: LCCN 2022061427 | ISBN 9781032444529 (hbk) | ISBN 9781032447896 (pbk) | ISBN 9781003373919 (ebk)
Subjects: LCSH: Building information modeling—Developing countries.
Classification: LCC TH438.13 .A34 2023 | DDC 690.01/13091724—dc23/eng/20230113 LC record available at https://lccn.loc.gov/2022061427

ISBN: 978-1-032-44452-9 (hbk)
ISBN: 978-1-032-44789-6 (pbk)
ISBN: 978-1-003-37391-9 (ebk)

DOI: 10.1201/9781003373919

Typeset in Times New Roman
by codeMantra

Contents

Part I

1   General Introduction

The construction industry is known to be fragmented in nature. The construction process consists of eight phases: strategic definition, preparation and brief, concept design, developed design, technical design, construction, handover and closeout, and in use (RIBA, 2013). The execution of these phases involves the expertise of different professionals and stakeholders. The fragmented nature of the construction industry requires a large amount of information to be exchanged among the construction parties. Each stage relies on the quality of information and the preceding phase. Therefore, a construction project's successful execution depends on the proper coordination of information to achieve the synergy of tasks and functions (Darshi de Saram & Ahmed, 2001). The successful management of information within project professionals ensures early clash detection and averts conflict. The resultant effect of a clash within a project team includes delay, which hampers project progress; cost and time overrun; friction; and an unhappy project team.

Considering the position of the construction industry as the cornerstone of the world economy (Castagnino et al., 2016), relevant tools for information management must be adopted. This is important to overcome the problems and achieve a more productive industry capable of meeting clients' needs. Technological tools possess the benefit of jumpstarting the process if harnessed by the construction industry. For instance, adopting new technologies improves completion time, quality, and safety and ensures an approximately 20% reduction in total lifecycle cost (Castagnino et al., 2016). However, compared to other industries, for instance, manufacturing, the construction industry is slow in technology adoption (Castagnino et al., 2016). Thus, it is not as productive as it should be. Hence, digitisation of the construction processes and phases is required to achieve an optimal output level in the construction industry. Adopting a technological tool capable of managing information within project stakeholders and phases is imperative.

The failure of the construction industry to adopt technological tools that will result in the fourth industrial revolution has been an uphill task. A viable and effective tool for collaboration and effective information management is building information modelling (BIM). BIM is a collaborative tool that cuts across all stakeholders in the construction industry (Eastman, 2011). It is an information management methodology that presents a standardised approach to information management (UKBIM alliance et al., 2019). BIM presents a frontier for collaboration against the

DOI: 10.1201/9781003373919-2

perceived norm of isolation in the construction industry. Also, it alters the existing fabric in terms of law, contract, organisational culture, and workflow, among others (RICS guidance note, 2014). The benefits of BIM transcend the professional barrier and lifecycle phase in the construction industry.

The emphasis of BIM is the effective management of information (UKBIM alliance et al., 2019), and it possesses a revolutionary power to transform the construction industry. Despite the inherent benefits of BIM, achieving full adoption across the industry is the major challenge (Gerbert et al., 2016). This is partly because the adoption of BIM has been lopsided along with professionals and geographical divides.

1.1   Objectives of the Book

Several studies have studied BIM adoption widely (Adekunle, Aigbavboa, Ejohwomu, Adekunle, et al., 2021; Adekunle, Ejohwomu, et al., 2021; Hamma-adama et al., 2018; Lindlar, 2014; Succar, 2008). It has been studied, and the benefits to the construction industry are severally established in the literature (Bryde et al., 2013; Eastman et al., 2008; Li et al., 2014). Nevertheless, the adoption has been skewed to the developed nation. For instance, according to Alufohai (2012), BIM adoption in Nigeria is predominantly among architects. The architects employ it mainly for aesthetics and not collaboratively with other professionals. Thus, it can be said that the architects are still at level zero according to the UK BIM wedge (BIM+, 2019); however, unlike others, they have embraced some BIM tools earlier than other stakeholders in the construction industry. Other professionals, for instance, quantity surveyors, are still struggling with the adoption although the awareness level is increasing (Adekunle, Aigbavboa, Ejohwomu, Efiannayi et al. 2022; Adekunle, Aigbavboa, Ejohwomu et al., 2021). An overview of BIM adoption in terms of geographical distribution reveals that BIM adoption is more among developed countries. Out of the six continents, Africa and the Middle East are considered beginners in BIM adoption (Jung & Lee, 2015). The developed countries where BIM has been adopted exhibited BIM leadership. To achieve this, they published different standards, regulations, legislations, and guides suited to their region. These initiatives are termed precursors for BIM implementation in organisations (Succar, 2010). These countries also adopted different approaches, but mostly a top-bottom approach has been said to be more effective (Troiani et al., 2020). This means that the government was the major driver of BIM adoption through legislation, for instance, in the United Kingdom.

However, developing countries are still struggling with BIM adoption. For instance, the construction industries in Nigeria and South Africa are still predominantly using traditional routes (Ogwueleka & Ikediashi, 2017). This corroborates the industry report by Autodesk (2005) in the South African construction industry. However, the study by Autodesk gave a promising prospect of BIM adoption. Comparing the South African construction industry to countries where BIM has been adopted, it is less mature, and there is a low rate of adoption (Chimhundu, 2015). A maturity model is required to provide a systematic, progressive, uniform assessment and framework for BIM adoption in developing countries. A maturity model puts all stakeholders on the same benchmark and provides a systematic adoption process framework. This study, therefore, developed a BIM maturity model for developing countries. The study contributed to the adoption of technology in the construction industry and, by extension, its effectiveness. The study reviewed selected existing BIM maturity models and gathered data through a mixed-method approach, and the analysed data were employed in developing the model to achieve this aim. This chapter and the subsequent ones chronicled the systematic development of the model.

1.2   Contribution and Value

This study was embarked upon to develop a BIM maturity model at a macro level for the South African construction industry. The contribution of this study is both theoretical and practical. Theoretically, this study is significant as it covers a gap in literature left unattended to by researchers in the South African construction industry and, in extension, developing countries. In the South African space, previous studies were focused on the factors affecting the adoption of BIM. However, none attempted to provide a maturity model for BIM. Thus, this study is focused on providing a systematic solution to BIM implementation through the developed maturity model.

The nation stands to gain immensely as an effective construction industry translates to a better economy for the industry. According to Gerbert et al. (2016), the construction industry is the cornerstone of the economy. A working construction industry invariably translates into a functional economy. A maturity model will provide the required framework for full BIM adoption in the South African construction industry. An efficient construction industry heralds international collaborations as all barriers due to non-conformance are broken down, thus many foreign investments and opportunities for across-border project collaborations. Besides, the implementation of BIM in construction reduces or eliminates project abandonment, non-performance, and reworks (Newton & Chileshe, 2012). Infrastructural development and a more effective built environment are the evidential significance of the book in the construction industry in developing countries.

1.3   Structure of the Book

This book is divided into four parts consisting of nine chapters. Part I provides the background to the book. In this part, the background to the book discusses an overview of the construction industry and its BIM potentials. It also presents the book's objectives and its contributions. Part II consists of chapters dwelling on BIM and its diffusion in the construction industry in developing countries. It also provides a literature review of the existing BIM adoption strategies globally. Lastly, it offers an insight into BIM as a driver for the optimisation of the construction process. Part III consists of two chapters focusing on defining a maturity model, the various types and characteristics of a maturity model, and the theoretical justification.

The other chapters reviewed existing maturity models extensively. The final part (Part IV) provides the conceptual perspective of the BIM maturity model and the conclusion. This part consists of three chapters (Chapters 7, 8 and 9). It explores the conceptual perspective of a BIM maturity model in a developing country. Each chapter has an introduction and summary of the chapter content.

1.4 Summary

This book introduces readers to the conceptual idea behind the book. It provides readers with the background to the book and its benefits and values. This chapter explores the skewed nature of BIM adoption in the construction industry and the need to achieve BIM adoption in developing countries. It, therefore, provides the need for a BIM maturity model for developing countries. It provides an insight into the gap to be filled by this book. The next chapter dwells on BIM diffusion in developing countries.

References

Adekunle, S. A., Aigbavboa, C. O., Ejohwomu, O., Adekunle, E. A., & Thwala, W. D. (2021). Digital Transformation in the Construction Industry: A Bibliometric Review. *Journal of Engineering, Design and Technology, 2013.* https://doi.org/10.1108/JEDT-08-2021-0442

Adekunle, S. A., Aigbavboa, C. O., Ejohwomu, O. A., Thwala, W. D., & Efiannayi, N. (2021). Key Constraints to Optimal BIM Penetration Among Nigerian Quantity Surveyors. In F. Emuze (Ed.), *Department of construction management 50th anniversary conference "the next 50 years" 15-16 November 2021, Port Elizabeth, South Africa* (pp. 66–73). https://pure.manchester.ac.uk/ws/portalfiles/portal/206816465/CM50_paper_22.doc.pdf

Adekunle, S., Aigbavboa, C., Ejohwomu, O., Efiannayi, N., John, B., & Thwala, W. (2022). Quantity Surveyors Scorecard in the 4IR: Unravelling the Bim Responsiveness in Developing Countries. *2022 European Conference on Computing in Construction Ixia, Rhodes, Greece* July 24–26, 2022, 4–9.

Adekunle, S. A., Aigbavboa, O. C., Ejohwomu, O. A., Oyeyipo, O., & Thwala, W. D. (2022). Unravelling the Encumberances to Better Information Management Among Quantity Surveyors in the 4IR: A Qualitative Study. In S. M. Ahmed, S. Azhar, A. D. Saul, & K. L. Mahaffy (Eds.), *12th International Conference on Construction in the 21st Century (CITC-12)* (pp. 105–111).

Adekunle, S. A., Ejohwomu, O., & Aigbavboa, C. O. (2021). Building Information Modelling Diffusion Research in Developing Countries: A User Meta-Model Approach. *Buildings, 11*(7), 264. https://doi.org/10.3390/buildings11070264

Alufohai, A. J. (2012). Adoption of Building Information Modelling and Nigeria's Quest for Project Cost Management. *FIG Working Week 2012 Knowing to Manage the Territory, Protect the Environment, Evaluate the Cultural Heritage,* 1–7.

Autodesk. (2005). *Autodesk ® Revit ® Building Information Modelling.* www.a3-inc.co.za

BIM+. (2019). *Explaining the Levels of BIM.* DIigital Construction Resource. http://www.bimplus.co.uk/analysis/explaining-levels-bim/

Bryde, D., Broquetas, M., & Volm, J. M. (2013). The Project Benefits of Building Information Modelling (BIM). *International Journal of Project Management, 2013.* https://doi.org/10.1016/j.ijproman.2012.12.001

Castagnino, S., Rothballer, C., & Gerbert, P. (2016). What's the Future of the Construction Industry? *World Economic Forum*. https://www.weforum.org/agenda/2016/04/building-in-the-fourth-industrial-revolution/

Chimhundu, S. (2015). *A Study on the BIM Adoption Readiness and Possible Mandatory Initiatives for Successful Implementation in South Africa*. University of the Witwatersrand, Johannesburg.

Darshi de Saram, B. D., & Ahmed, S. M. (2001). *Construction Coordination Activities: What Is Important and What Consumes Time. Journal of Management in Engineering*, *17*(4), 203–213.

Eastman, C. M. (2011). *BIM Handbook : A Guide to Building Information Modeling for Owners, Managers, Designers, Engineers and Contractors*. Wiley. https://books.google.co.za/books?hl=en&lr=&id=aCi7Ozwkoj0C&oi=fnd&pg=PP7&ots=ZbDeOUz7Iq&sig=yvlZ_9-KVeVMEAmjXEog94FzI7Q&redir_esc=y#v=onepage&q&f=false

Eastman, C., Teicholz, P., Sacks, R., & Liston, K. (2008). *BIM Handbook A Guide to Building Information Modeling for Owners, Managers, Designers, Engineers, and Contractors*. John Wiley & Sons, Inc. https://doi.org/10.1093/nq/s7-II.32.110-e

Gerbert, P., Castagnino, S., Rothballer, C., Renz, A., & Filitz, R. (2016). *The Transformative Power of Building Information Modeling*. https://www.bcg.com/publications/2016/engineered-products-infrastructure-digital-transformative-power-building-information-modeling.aspx

Hamma-adama, M., Salman, H., & Kouider, T. (2018). Diffusion of Innovations: The Status of Building Information Modelling Uptake in Nigeria. *Journal of Scientific Research and Reports*, *17*(4), 1–12. https://doi.org/10.9734/jsrr/2017/38711

Jung, W., & Lee, G. (2015). The Status of BIM Adoption on Six Continents. *International Journal of Civil and Environmental Engineering*, *9*(5), 512–516. https://www.semanticscholar.org/paper/The-Status-of-BIM-Adoption-on-Six-Continents-Jung-Lee/ea0d7a32ebe25d64509e4224e6be9371c6aa1369#paper-header

Li, J., Wang, Y., Wang, X., Luo, H., Kang, S.-C., Wang, J., Guo, J., & Jiao, Y. (2014). Benefits of Building Information Modelling in the Project Lifecycle: Construction Projects in Asia. *International Journal of Advanced Robotic Systems*, *11*(8), 124. https://doi.org/10.5772/58447

Lindlar, M. (2014). Building Information Modeling – A Game Changer for Interoperability and a Chance for Digital Preservation of Architectural Data? *Proceedings of the 11th International Conference on Preservation of Digital Objects: Proceedings - IPRES 2014 - Melbourne*. https://doi.org/11353/10.378117

Newton, K., & Chileshe, N. (2012). Awareness, Usage and Benefits of Building Information Modelling (BIM) Adoption – The Case of the South Australian Construction Organisations. In S.D. Smith, (Ed.), *Procs 28th Annual ARCOM Conference, 3–5 September 2012, Edinburgh, UK, Association of Researchers in Construction Management, 3–12* (Vol. 02, Issue May, pp. 3–12). https://doi.org/10.13140/RG.2.1.2352.3363

Ogwueleka, A. C., & Ikediashi, D. I. (2017). The Future of BIM Technologies in Africa: Prospects and Challenges. In *Integrated Building Information Modelling* (Issue July, pp. 307–314). https://doi.org/10.2174/9781681084572117010016

RIBA. (2013). *RIBA Plan of Work 2013 Overview*. www.ribaplanofwork.com

RICS guidance note. (2014). *International BIM Implementation Guide* (Issue September).

Succar, B. (2008). Building Information Modelling Framework: A Research and Delivery Foundation for Industry Stakeholders. *Automation in Construction*, *18*, 357–375. https://doi.org/10.1016/j.autcon.2008.10.003

Succar, B. (2010). The Five Components of BIM Performance Measurement. *Proceedings of CIB World Congress, Salford*. https://doi.org/10.1136/bmj.3.5560.312-a

Troiani, E., Mahamadu, A.-M., Manu, P., Kissi, E., Aigbavboa, C., & Oti, A. (2020). Macro-maturity Factors and Their Influence on Micro-Level BIM Implementation within Design Firms in Italy. *Architectural Engineering and Design Management*. https://doi.org/10.1080/17452007.2020.1738994

UKBIM alliance, CDBB, & BSI. (2019). *Information Management According to BS EN ISO 19650*.

Part II

2 Understanding BIM Diffusion in the Construction Industry in Developing Countries

2.1 Introduction

Over the years, BIM adoption has been observed to be skewed. It is more advanced and widely adopted in developed countries. Developing countries are still struggling to achieve BIM adoption due to several barriers. This chapter focuses on developing countries in the context of BIM diffusion. It identifies the challenges of the construction industry in developing countries and the barriers to BIM adoption. Finally, it adopts the South African construction industry (SACI) as a case laboratory for understanding BIM implementation in the global south.

2.2 Overview of the Construction Industry in Developing Countries

The construction industry in developing countries has been severally studied from different perspectives and contexts. To understand the construction industry in developing countries, Boadu et al. (2020) conducted a study on the Ghanaian construction industry. The study findings revealed three critical challenges affecting the construction industry in developing countries focusing on health and safety performance. It was revealed that there is a lack of skilled and educated workforce, a reliance on labour-intensive methods, and a lack of a single regulatory authority. It is evident that developing countries are challenged by a lack of an effective regulatory framework.

On the contrary, not having a single authority means there is a plethora of regulations, and regulations are not standardised; hence, the diverse regulations translate to difficulty in adequately regulating the construction industry. Also, it might translate to a clash of the existing regulatory frameworks. Thus, there is a need to harmonise and implement regulations.

Another study by Moavenzadeh (1978) on developing countries looked at the issues during the design and construction phases. The study identified a dearth of experienced manpower, a lack of adequate credit facilities, immobility of the labour force, and lack of maintenance, among others. The challenges can effectively be classified into manpower, financial/ease of doing business, and regulatory challenges, among others. Other challenges in the industry are project performance

DOI: 10.1201/9781003373919-4

related. Adekunle et al. (2020) posit that payment issues, construction difficulty, and design issues are the top three factors causing a lag in project performance in developing countries. This suggests that project designs by professionals and payment-related issues are impacting project delivery.

Another notable study articulating the challenges of the construction industry in developing countries is the study by (Ofori, 2000). The study posits that the construction industry in developing countries is not dynamic with the changing times; hence, they stick to borrowed processes and cultures that have been reviewed in the originating countries. Thus, aside from the manpower, education, and financial/economic challenges, the construction industry in developing countries is plagued with rigidity to change. The construction industries stick to old processes that are outdated in the event of new realities.

New realities in this context refer to the fourth industrial revolution and its emerging technologies. These emerging technologies are important to the construction industry and are constantly changing the process, product, and people (Adekunle, Aigbavboa et al., 2021; Ejohwomu et al., 2021). Sticking to the status quo will, therefore, not achieve a transformation of the industry. Unfortunately, the reality in developing countries is that they stick to the status quo; hence, there is a need to help the industry achieve new, proven, and effective ways of the construction industry like the developed countries. Several solutions have been suggested to help the construction industry in overcoming the identified challenges; these include proper manpower planning (Uwakweh & Maloney, 1991), a proper study of the culture and environment (Ofori, 2000), and multi-national construction firms as a potential conduit for the transformation of the construction industry in developing countries (Moavenzadeh & Rossow, 1975). However, there is a need for a stepwise solution to transform the construction industry in developing countries.

BIM diffusion research in developing countries reveals a promising but very low diffusion rate as compared to developed countries. A study by Adekunle, Ejohwomu et al. (2021) revealed that BIM diffusion research is most in Taiwan, followed by Malaysia, Brazil, and India. It was also revealed that BIM diffusion research started in 2006 in developing countries as opposed to 2004, when the first publication was observed in developed countries (Liu et al., 2019). Despite this observation and the proximity of the commencement of BIM research, BIM diffusion is wide apart in both contexts.

In the study by Liu et al. (2019) on global BIM research, it was observed that the top five productive countries are the United States, China, England, Australia, and South Korea. This aligns with the study by Jung and Lee (2015) where the BIM adoption of the six continents was studied and compared. It was observed that North America, Oceania, and Europe ranked higher than other continents.

2.3   The South African Construction Industry (SACI) as a Case Study

The industry is plagued with project cost overrun, payment difficulties, lack of effective communication on a project (Dithebe et al., 2018), and slim profit margins by contractors (Harris, 2019), among other challenges. However, the technology was perceived as low ranking in the challenges facing the SACI (Windapo & Cattell, 2013). It has been proven that technology adoption is central to the transformation of the construction industry (Gerbert et al., 2016; WEF & BCG, 2016). Technology impacts every phase of the industry and removes unnecessary downtimes and bottlenecks in the SACI. Hence, most of the identified problems facing the SACI and inhibiting its performance can be solved by digitising the industry. In their study, digitisation for effective project delivery in SACI, Aghimien, Aigbavboa and Oke (2018) opined that digitisation in SACI must be strategic in order to achieve a more productive and efficient construction industry.

Despite the centrality of digitisation, the research outputs focusing on SACI have been observed to be few. Some of the works on digitisation in the SACI are:

Reference	Title
Aghimien , Aigbavboa, Oke, and Koloko (2018)	Digitalisation in the South African construction industry: Construction professionals perspective
Oke et al. (2018)	Challenges of digital collaboration in the South African construction industry

In the two studies identified above, professionals opined that digital technologies are considerably being used in the SACI, albeit more at the design stage of construction projects. Also, it is worthy of note that digital technologies are in use. Furthermore, professionals find it hard to collaborate digitally due to a lack of training, the cost of implementing the technologies, and interoperability. The identified studies are generic in nature; this chapter is specific in its study by critically analysing BIM maturity in SACI.

2.4   Building Information Modelling and the South African Construction Industry

The SACI has been observed to have a low maturity in BIM implementation (Akintola, Root, & Venkatachalam 2017) and has been adopted on large projects (Kiprotich, 2014). Although there have been different studies and efforts focused on the widespread adoption of BIM in SACI, the industry is yet to achieve this across all stakeholders. According to Chimhundu (2015), some level of BIM was adopted during the massive infrastructural demand due to the 2010 FIFA™ World Cup hosted by South Africa. However, this did not achieve widespread adoption among stakeholders. Majorly, BIM has been adopted for design functions and sparsely by professionals for collaboration with contractors and scheduling in SACI (Froise & Shakantu, 2014). A systematic search of the literature reveals Table 2.1 as the BIM publications in the SACI. A critical analysis of these publications reveals that BIM feasibility in the SACI can be classified into barriers to implementation, impact on professionals, and new actor BIM roles in SACI.

Table 2.1 BIM publications in South Africa

References	Title
Harris (2019)	BIM: it is your move
Akintola et al. (2016)	The impact of implementing BIM on AEC organisational workflows
Akintola, Venkatachalam and Root (2017)	New BIM roles' legitimacy and changing power dynamics on BIM-enabled projects
Beukes (2012)	How a quantity surveyor in South Africa can use building information modelling (BIM) to stay relevant in the construction industry
Kekana et al. (2014)	Building information modelling (BIM): Barriers in adoption and implementation strategies in the South Africa construction industry
Chimhundu (2015)	A study on the BIM adoption readiness and possible mandatory initiatives for successful implementation in South Africa
Froise and Shakantu (2014)	Diffusion of innovations: An assessment of building information modelling uptake trends in South Africa
Kaber (2010)	Will the implementation of building information modelling be advantageous to the South African construction industry?
Monyane and Ramabodu (2014)	Exploration of building information modelling (BIM) concept and its effects on quantity surveying profession in South Africa: Case of FS Province
Froise (2014)	Building information modelling as a catalyst for an integrated construction project delivery culture in South Africa
Kekana et al. (2015)	Understanding building information modelling in the South Africa construction industry
Odubiyi et al. (2019)	Strategies for building information modelling adoption in the South African construction industry
Ogwueleka and Ikediashi (2017)	The future of BIM technologies in Africa: Prospects and challenges
Kiprotich (2014)	An investigation on building information modelling in project management: Challenges, strategies and prospects in the Gauteng construction industry, South Africa
Akintola, Root and Venkatachalam (2017)	Key constraints to optimal and widespread implementation of BIM in the South African construction industry
Pillay et al. (2018)	Use of BIM at higher learning institutions: Evaluating the level of implementation and development of BIM at built environment schools in South Africa
Olugboyega & Windapo, (2021)	Modelling the indicators of a reduction in BIM adoption barriers in a developing country
Olugboyega & Windapo, (2021b)	Structural equation model of the barriers to preliminary and sustained BIM adoption in a developing country
Moodley et al. (2016)	Teaching BIM in schools of architecture of South African universities
Akintola et al. (2020)	Understanding BIM's impact on professional work practices using activity theory

2.5 Barriers to BIM Implementation in the South African Construction Industry

The SACI has been confronted with different challenges when implementing BIM; these challenges are presented in Table 2.2. A table analysis of these barriers shows that the lack of interest by stakeholders is the major barrier to BIM implementation in the SACI. This might be connected to the lack of BIM awareness – the third most mentioned barrier in literature. It can be inferred that the lack of interest by stakeholders stems from the lack of awareness. Hence, if stakeholders are made aware of the benefits of BIM implementation, they might be a substantial adoption of BIM in the SACI.

The second barrier commonly identified in the literature is jointly found in the absence of BIM specialists and the absence of a legal framework supporting the implementation of BIM. Although the absence of BIM professionals might be traced to a lack of BIM education and training, the BIM process has birthed new roles beyond just training that adds skills. The BIM actors are new roles required for the seamless implementation of BIM, although these roles are yet to gain legitimacy through wide acceptance. Meanwhile, these roles are currently without a standardised training framework and are mostly performed by established industry professionals with acquired skills (Akintola et al., 2020). Regarding the absence of a legal framework, the SACI requires a set of standards that recognises and details the new actor roles, the work breakdown structure, and level of detail, among others peculiar to the BIM process on construction projects. The existing frameworks do not recognise these aspects, hence the challenge of the implementation of BIM using the existing standards.

2.5.1 The Necessity of a BIM Maturity Model in the SACI

From the literature, it has been established that there is a challenge with the widespread implementation of BIM in the SACI as well as in other developing countries. To achieve a globally competitive and digitally matured construction industry, the SACI must overcome these challenges. Therefore, a systematic, methodological, and stepwise framework is required to assist all stakeholders with BIM implementation. Beyond the government mandates, incentives, and awareness, there must be an intentional guide to assist stakeholders through implementation to achieve full maturity.

Meanwhile, there are existing maturity models that have been adopted in other countries that could easily be adopted in South Africa. However, it should be noted that maturity models are not one size fits all (Kassem et al., 2020). BIM maturity models respond to the prevailing economic, social, and cultural, among other, conditions in a country. Hence, the need for a BIM maturity model to assist stakeholders from the present low implementation level and guide them in a stepwise process to full BIM maturity. The BIM maturity model specifically developed for the SACI will achieve the missing link in the BIM implementation efforts. This will complement other efforts by the different stakeholders and provide a stepwise implementation pathway.

Table 2.2 Barriers to BIM adoption in SACI

Barriers	Harris (2019)	Akintola, Root and Venkatachalam (2017)	Akintola et al. (2016)	Kekana et al. (2014, 2015)	Monyane and Ramabodu (2014)	Chimhundu (2015)	Kiprotich (2014)
Lack of local demand, interest, and support for BIM by stakeholders (Asset Managers, government, clients, consultant, and owner operators)	x	x				x	x
Lack of skilled BIM practitioners	x						x
Insufficient awareness		x		x			x
Absence of supportive legislations and standards (legal and contractual framework).		x		x			x
Resistance to change			x			x	
High set-up and training cost			x				
Lack of BIM experience and education						x	
Not enough internal resource					x		
Level of training required					x		
Lack of technical support for interoperability							x
Lack of BIM's capability							x

2.6   Summary

This chapter described the BIM implementation from the developing country perspective. It explored existing studies on the construction industry in developing countries to identify its peculiarities and challenges in order to provide insight into the construction industry in the context. The study also explored BIM implementation as it affects the SACI. It analysed the barriers to implementation and identified existing BIM studies in the South African context. The chapter thus provided a holistic insight into BIM implementation in developing countries and identified the peculiarities of the industry in developing countries.

References

Adekunle, S. A., Aigbavboa, C., & Ejohwomu, O. A. (2020). Improving Construction Project Performance in Developing Countries: Contractor Approach. *3rd European and Mediterranean Structural Engineering and Construction Conference 2020, Euro-Med-Sec 2020, 7*(1). https://doi.org/10.14455/isec.res.2020.7(1).con-14

Adekunle, S. A., Aigbavboa, C. O., Ejohwomu, O., Adekunle, E. A., & Thwala, W. D. (2021). Digital Transformation in the Construction Industry : A Bibliometric Review. *Journal of Engineering, Design and Technology, 2013.* https://doi.org/10.1108/JEDT-08-2021-0442

Adekunle, S. A., Ejohwomu, O., & Aigbavboa, C. O. (2021). Building Information Modelling Diffusion Research in Developing Countries: A User Meta-Model Approach. *Buildings, 11*(7), 264. https://doi.org/10.3390/buildings11070264

Aghimien, D. O., Aigbavboa, C. O., & Oke, A. (2018). Digitisation for Effective Construction Project Delivery in South Africa: A Review. *Contemporary Construction Conference: Dynamic and Innovative Built Environment (CCC2018)At: Coventry, United Kingdom.*

Aghimien, D., Aigbavboa, C., Oke, A., & Koloko, N. (2018). Digitalisation in the South African Construction Industry : Construction Professionals Perspective. *Proceedings of the Fourth Australasia and South-East Asia Structural Engineering and Construction Conference At: Brisbane, Australia.*

Akintola, A., Douman, D., Kleynhans, M., & Maneli, S. (2016). The Impact of Implementing BIM on AEC Organisational Workflows. *9th CIDB Postgraduate Conference, Emerging Trends in Construction Organisational Practices and Project Management Knowledge Area, February 2–4, 2016, Cape Town, South Africa.*, 506–516.

Akintola, A., Root, D., & Venkatachalam, S. (2017). Key Constraints to Optimal and Widespread Implementation of BIM in the South African Construction Industry. *Proceeding of the 33rd Annual ARCOM Conference, 4–6 September 2017*, 25–34. http://www.arcom.ac.uk/-docs/proceedings/fd7c938fcee1ffc085f098c3b88f8c34.pdf

Akintola, A., Venkatachalam, S., & Root, D. (2017). New BIM Roles' Legitimacy and Changing Power Dynamics on BIM-Enabled Projects. *Journal of Construction Engineering and Management, 143*(9). https://doi.org/10.1061/(ASCE)CO.1943-7862.0001366

Akintola, A., Venkatachalam, S., & Root, D. (2020). Understanding BIM's Impact on Professional Work Practices Using Activity Theory. *Construction Management and Economics, 38*(5), 447–467. https://doi.org/10.1080/01446193.2018.1559338

Beukes, D. S. (2012). *How a Quantity Surveyor in South Africa can use Building Information Modeling (BIM) to Stay Relevant in the Construction Industry.* University of Pretoria.

Boadu, E. F., Wang, C. C., & Sunindijo, R. Y. (2020). Characteristics of the Construction Industry in Developing Countries and Its Implications for Health and Safety: An Exploratory

Study in Ghana. *International Journal of Environmental Research and Public Health*, *17*(11), 4110. https://doi.org/10.3390/IJERPH17114110

Chimhundu, S. (2015). *A Study on the BIM Adoption Readiness and Possible Mandatory Initiatives for Successful Implementation in South Africa*. University of the Witwatersrand, Johannesburg.

Dithebe, K., Aigbavboa, C., Oke, A., & Muyambu, M. A. (2018). Factors Influencing the Performance of the South African Construction Industry: A Case of Limpopo Province. *Proceedings of the International Conference on Industrial Engineering and Operations Management*, 1185–1192.

Ejohwomu, O., Adekunle, S. A., Aigbavboa, O. C., & Bukoye, T. (2021). Construction and Fourth Industrial Revolution: Issues and Strategies. In P. Emmanuel Adinyira, K. Agyekum (Eds.), *The Construction Industry: Global Trends, Job Burnout and Safety Issues* (First). Nova Science Publishers. https://doi.org/10.52305/JDFM1229

Froise, T. (2014). *Building Information Modelling as a Catalyst for an Integrated Construction Project Delivery Culture in South Africa* (Issue March). https://www.academia.edu/10450016/BIM_as_a_Catalyst_for_an_Integrated_Construction_Project_Delivery_Culture_in_South_Africa

Froise, T., & Shakantu, W. (2014). Diffusion of Innovations: an Assessment of Building Information Modelling Uptake Trends in South Africa. *Journal of Construction Project Management and Innovation*, *4*(2), 895–911.

Gerbert, P., Castagnino, S., Rothballer, C., Renz, A., & Filitz, R. (2016). *The Transformative Power of Building Information Modeling*. https://www.bcg.com/publications/2016/engineered-products-infrastructure-digital-transformative-power-building-information-modeling.aspx

Harris, V. (2019). *BIM-It's Your Move*. www.bimacademyafrica.co.za

Jung, W., & Lee, G. (2015). The Status of BIM Adoption on Six Continents. *International Journal of Civil and Environmental Engineering*, *9*(5), 512–516. https://www.semanticscholar.org/paper/The-Status-of-BIM-Adoption-on-Six-Continents-Jung-Lee/ea0d7a32ebe25d64509e4224e6be9371c6aa1369#paper-header

Kaber, R. (2010). *Will the Implementation of Building Information Modelling Be Advantagous to the South African Construction Industry?* University of Pretoria.

Kassem, M., Li, J., Kumar, B., Malleson, A., Gibbs, D. J., Kelly, G., & Watson, R. (2020). *Building Information Modelling: Evaluating Tools for Maturity and Benefits Measurement*. Centre for Digital Built Britain, 184.Kekana, T., Aigbavboa, C., & Thwala, W. (2014). Building Information Modelling (BIM): Barriers in Adoption and Implementation Strategies in the South Africa Construction Industry. *International Conference on Emerging Trends in Computer and Image Processing, December 15–16*.

Kekana, G., Aigbavboa, C., & Thwala, W. (2015). Understanding Building Information Modelling in the South Africa Construction Industry. *12th International OTMC Conference: Organisation, Technology and Management in Construction, September 2015*.

Kiprotich, C. J. (2014). *An Investigation on Building Information Modelling in Project Management: Challenges, Strategies and Prospects in the Gauteng*. University of the Witwatersrand.

Liu, Z., Lu, Y. & Peh, L. C. (2019). A Review and Scientometric Analysis of Global Building Information Modeling (BIM) Research in the Architecture, Engineering and Construction (AEC) Industry. *Buildings*, *9*(10), 210. https://doi.org/10.3390/buildings9100210

Moavenzadeh, F. (1978). Construction Industry in Developing Countries. *World Development*, *6*(1), 97–116. https://doi.org/10.1016/0305-750X(78)90027-X

Moavenzadeh, F., & Rossow, K. J. A. (1975). *The Construction Industry in Developing Countries Professor of Civil Engineering*. Massachusetts Institute of Technology.

Monyane, T. G., & Ramabodu, M. S. (2014). Exploration of Building Information Modelling (BIM) Concept and Its Effects on Quantity Surveying Profession in South Africa: Case of FS Province. *7th Annual SACQSP Research Conference on "Mapping the Future," 0035*, 419–429. https://www.researchgate.net/profile/Thabiso_Godfrey_Monyane/publication/273888472_Exploration_of_Building_Information_Modelling_BIM_concept_and_its_eff_ects_on_quantity_surveying_profession_in_South_Africa_case_of_FS_Province/links/550fe31a0cf21287416c5b

Moodley, V., Mathye, K., & Radebe, S. (2016). *Teaching BIM in Schools of Architecture of South African Universities*. University of the Witwatersrand.

Odubiyi, T. B., Aigbavboa, C., Thwala, W., & Netshidane, N. (2019). Strategies for Building Information Modelling Adoption in the South African Construction Industry. *Modular and Offsite Construction (MOC) Summit Proceedings*, 514–519. https://doi.org/10.29173/mocs133

Ofori, G. (2000). Challenges of Construction Industries in Developing Countries: Lessons from Various Countries. *2nd International Conference on Construction in Developing Countries: Challenges Facing the Construction Industry in Developing Countries, Gaborone*, 15–17.

Ogwueleka, A. C., & Ikediashi, D. I. (2017). The Future of BIM Technologies in Africa: Prospects and Challenges. In *Integrated Building Information Modelling* (Issue July, pp. 307–314). https://doi.org/10.2174/9781681084572117010016

Oke, A. E., Aghimien, D. O., Aigbavboa, C. O., & Nteboheng, K. (2018). Challenges of Digital Collaboration in The South African Construction Industry. *Proceedings of the International Conference on Industrial Engineering and Operations Management Bandung, Indonesia, March 6–8, 2018*, 2472–2482.

Pillay, N., Musonda, I., & Makabate, C. (2018). Use of BIM at Higher Learning Institutions : Evaluating the Level of Implementation and Development of BIM at Built Environment Schools in South Africa. *Aubea 2018- Educating Building Professionals for Thr Future: Innovation, Technology, Sustainability. Singapore Volume: 2, 2*, 227–240.

Uwakweh, B. O., & Maloney, W. F. (1991). Construction Management and Economics Conceptual Model for Manpower Planning for the Construction Industry in Developing Countries. *Construction Management and Economics*, *9*(5), 451–465. https://doi.org/10.1080/01446199100000034

WEF, & BCG. (2016). *Shaping the Future of Construction A Breakthrough in Mindset and Technology*. World Economic Forum.

Windapo, A. O., & Cattell, K. (2013). The South African Construction Industry: Perceptions of Key Challenges Facing Its Performance, Development and Growth. *Journal of Construction in Developing Countries*, *18*(2), 65.

3 Global BIM Adoption Strategies

3.1 Introduction

The adoption of building information modelling (BIM) globally as a collaborative tool to improve the productivity and efficiency of the construction industry has taken the front burner in the past decade. Different countries adopted diverse strategies to achieve its diffusion. This is because there is no one-cap-fits-all strategy; different countries adopted strategies that are responsive to the prevailing conditions (Adekunle et al., 2022). This chapter presents an overview of the BIM adoption strategies globally. The adoption of BIM is more skewed towards developed countries; hence, the strategies discussed herein are focused on BIM-leading countries majorly.

3.2 Overview of BIM Adoption Globally

BIM is globally acclaimed and adopted for its transformative potential for the construction industry. To this end, it has a worldwide acceptance, and various construction industries around the world have attempted and adopted it. However, the adoption is not the same when observed globally. Thus, the adoption level is different globally.

There are different metrics for assessing BIM performance and penetration. Succar (2010) highlights five components: Capability, Maturity, Competency, Organisational Scale, and Granularity. These assessment metrics are designed to assess the level of BIM capability and maturity when applied. They consist of sub-metrics to measure different levels of BIM adoption. It also provides a systemic assessment approach.

The study by Jung and Lee (2015) adopted four indexes to assess BIM adoption across the continents globally. They adopted the engagement level, the Hype Cycle model, the technology diffusion model, and BIM services. The study succeeded in assessing BIM adoption through the adopters, the phase of adoption, and the BIM services frequently used across the continents. The two studies adopt a practical approach to assess BIM periodically based on the prevailing BIM adoption status in the industry of focus.

DOI: 10.1201/9781003373919-5

Jung and Lee (2015) presented the status of BIM adoption across the six continents. Although the study had a limitation of a few respondents, however, it provided insight into the adoption level and rated the six continents. The study observed that the most advanced continent is North America, followed by Oceania and Europe. South America ranks the lowest; meanwhile, the Middle East and Africa were described to be at the beginner phase.

A quick glance at Figure 3.1 shows the BIM adoption efforts and plans globally. Some of these adoptions are government-driven and support with mandates for adoption, for instance, in the United States, Peru, and Brazil, among others. The government plays a multifaceted role in the implementation of BIM beyond mandates. Most of these BIM-leading countries have dedicated BIM policy documents tailored to their context, for instance, Japan. Table 3.1 presents some standards introduced to foster BIM implementation; some of the standardisation bodies involved with this are the International Organisation for Standardisation (ISO) and the European Committee for Standardisation, among others. Most of these countries have a well-tailored BIM implementation plan and programme, e.g., Canada. This implementation programme assists in the systematic adoption of BIM, allocation of roles to stakeholders, and providing insights to stakeholders to achieve macro-level BIM diffusion. A critical study of the figure also shows that no African country has any policy, implementation plan, effort, or BIM footprint. This is similar to most developing countries, for instance, many are still struggling with awareness (Adekunle et al., 2020; Zaini et al., 2019). To understand BIM adoption in developing countries, Ismail et al. (2017) explored the BIM uptake in Asian developing countries. According to the study, some of the developing countries do not have any information on BIM implementation efforts and were dropped. Initially, the study selected Bangladesh, Cambodia, China, India, Indonesia, Laos, Malaysia, Maldives, Mongolia, Myanmar, Nepal, North Korea, Pakistan, Philippines, Sri Lanka, Thailand, and Vietnam. This was reduced to China, India, Malaysia, Indonesia, Thailand, Myanmar, Sri Lanka, Mongolia, Vietnam, and Pakistan. It was observed that most of these countries experienced slow adoption while some are yet to adopt it (e.g. Sri Lanka). Also, they are mostly still grappling with the barriers. However, a few have developed BIM roadmaps.

3.3   International Building Information Modelling Standard

Various standards have been developed and adopted to achieve faster diffusion and a regulated BIM implementation environment. These standards aid in the implementation governance of the BIM. This section discusses two major BIM standards in the construction industry: PAS 1192 and the ISO 19650 family of standards. These standards are focused on information management through the use of BIM throughout the building lifecycle.

The PAS 1192 has the following components BS 1192: 2007 (collaboration methodology for managing production, distribution, and the quality of building

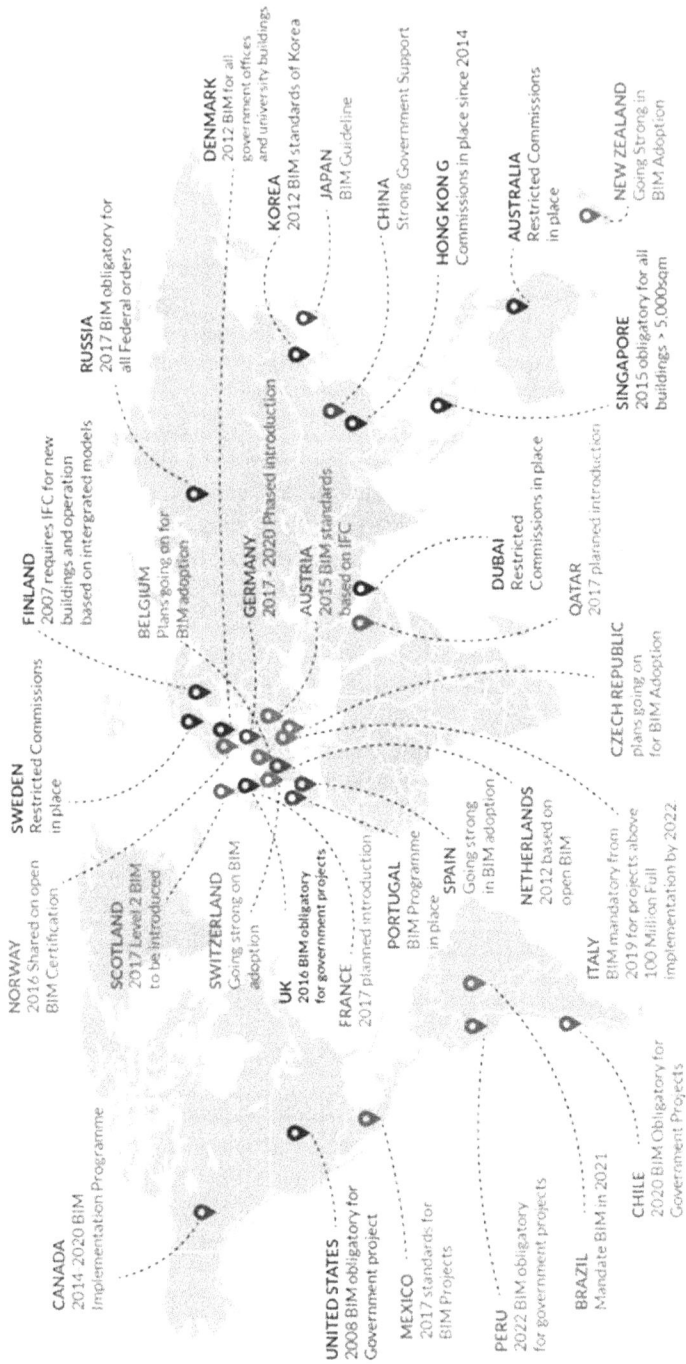

Figure 3.1 Global BIM adoption strategies (Techture ISO 9001-2015 Certified 2020)

Table 3.1 Some published BIM standards

Standard	Description
Organisation of information about construction works – Framework for management of project information ISO 22263	This standard outlines the organisation of project information (process and product-related) in construction projects.
Framework for building information modelling (BIM) guidance ISO/TS 12911	It provides a framework for commissioning building information modelling (BIM) specifications.
Guidelines for knowledge libraries and object libraries ISO 16354	It outlines the categories of knowledge libraries. It also lays the foundation for uniformity in terms of the structures and content of the knowledge libraries.
Data structures for electronic product catalogues for building services – Part 1: Concepts, architecture and model ISO 16757-1	This standard provides data structures for electronic product catalogues for the transmission of building services product data routinely into models of building services software applications.
Building construction – Organisation of information about construction works – Part 2: Framework for classification ISO 12006-2	It defines a framework for developing built environment classification systems.
Data structures for electronic product catalogues for building services – Part 2: Geometry ISO 16757-2	This standard describes the geometry for the modelling of building services products.
Industry Foundation Classes (IFC) for data sharing in the construction and facility management industries – Part 1: Data schema ISO 16739-1	This standard includes definitions covering the required data for buildings throughout their life cycle.
Building construction – Organisation of information about construction works – Part 3: Framework for object-oriented information EN ISO 12006-3	The standard specifies a language-independent information model that can be used for developing dictionaries for storing or providing information about construction works.
Building information models – Information delivery manual – Part 2: Interaction framework EN ISO 29481-2	It specifies a methodology and format to describe coordination roles between building project actors throughout its life cycle stages.
Building information models – Information delivery manual – Part 1: Methodology and format EN ISO 29481-1	This information delivery standard is focused on software applications interoperability.
Organisation and digitisation of information about buildings and civil engineering works, including building information modelling (BIM) – Information management using building EN ISO 19650-1 information modelling – Part 1: Concepts and principles	The document sets out the recommended concepts and principles for business processes across the built environment sector for information management during the life cycle of built assets

(Continued)

Table 3.1 (Continued)

Standard	Description
Organisation and digitisation of information about buildings and civil engineering works, including building information modelling (BIM) – Information management using building EN ISO 19650-2 information modelling – Part 2: Delivery phase of the assets	This document specifies requirements for information management during the delivery phase of assets using building information modelling.
Organisation and digitisation of information about buildings and civil engineering works, including building information modelling (BIM) – Information management using building information modelling – Part 5: Security-minded approach to information management (ISO 19650-5:2020)	The standard specifies the principles and requirements for security-minded information management for BIM

information), BS 1192-2 and BS 1192-3 (for level 2, explain and specify information management requirements), BS 1192-4 (this defines the use of COBie throughout facility lifecycle for information exchange), and BS 1192-5 (this addresses the security issues in the digital built environment. The PAS 1192-5 is a standard guiding BIM implementation, while PAS 1192-2 was produced to promote and ensure the information lean thinking throughout a project life cycle. According to British Standards Institution (BSI) (2013), the BS 1192 was developed to avoid wasteful activities on construction projects. These wasteful activities include waiting and searching for information, over-production of information without a defined use, technological use for over-processing information, and rework caused by defects. The PAS 1192 has been adopted in the development of BIM maturity models. For instance, it forms the basis for the UK BIM wedge till level 2.

This PAS 1192 family of standards has been withdrawn and replaced by ISO 19650. The new standard is focused on the information management capability of BIM and other related issues over the lifecycle of a built asset. Like the preceding standard, it has five components: ISO 19650-1 (Organisation and digitisation of information about buildings and civil engineering works, including building information modelling [BIM] – Information management using building information modelling): Concepts and principles; ISO 19650-2 (Organisation and digitisation of information about buildings and civil engineering works, including building information modelling [BIM] – Information management using building information modelling : Delivery phase of the assets); ISO 19650-3 (Organisation and digitisation of information about buildings and civil engineering works, including building information modelling [BIM] – Information management using building information modelling: Operational phase of the asset); ISO 19650-4 (Organisation and digitisation of information about buildings and civil engineering works, including building information modelling [BIM] – Information management using building information modelling), and ISO 19650-5:2020: (Organization and digitization of information about buildings and civil engineering works, including building information modelling (BIM) – Information management using building information modelling – Security-minded approach to information management).

3.4 Global BIM Implementation Strategies

BIM adoption has been unequal globally; thus, different countries exhibit different levels of adoption. Countries that are more advanced in the adoption level than others are referred to as BIM leaders. These countries adopted different strategies based on their environmental conditions – economic, infrastructural, and political – because there is no one-size-fits-all strategy. BIM is widely used in the United States, the United Kingdom, France, Germany, Finland, Denmark, Australia, Malaysia, and Singapore (Kubba, 2017). This section discusses various implementation strategies adopted globally in some selected countries.

3.4.1 Australia

The BIM initiative was taken by BEIIC (Built Environment Industry Innovation Council), which was named the National Building Information Modelling (NBIM). According to Autodesk, ConsultAustralia and AIA (2010) and Edirisinghe and London (2015), six areas were identified for national action to achieve BIM adoption: procurement and legal issues, BIM guidelines, multidisciplinary BIM education, product information libraries, business process change, and compliance and certification issues as a matter of priority. Some initiatives in the Australian construction industry include the NATSPEC National BIM guide and the ACIF-APPC BIM framework. BIM adoption has been reportedly high and promising in Australia. Contractors in the Australian construction industry reported a 78% positive return on investment (ROI) (McGraw Hill, 2014).

3.4.2 United Kingdom

The United Kingdom is one of the BIM leaders. The government drove the adoption in the United Kingdom through mandates. For instance, the government mandated that all public sector projects must adopt BIM level 2 by 2016. Different studies have been conducted to provide a pathway for achieving the government mandates, one of which is Khosrowshahi and Arayici (2012). This study provided the framework for the BIM 5D adoption in the United Kingdom. Also, the popular Bew and Richards model was developed in the United Kingdom, and it assisted in the BIM adoption in the United Kingdom construction industry. The successful BIM application on real-life projects was at the Heathrow Terminal 5 project. It was reported that its adoption caused a reduction of 210m pounds in the project cost. Another initiative employed by the government was the creation of the BIM task group; the group was created to assist and support organisations in the transition to BIM.

3.4.3 United States

The United States is a leading BIM adopter and has experienced advanced research in this area. As of 2012, the United States boasted of more than 70% adoption rate, and a high ROI rate was also observed (McGraw Hill, 2012). It was reported that the major driver in the United States is similar to that of the United Kingdom, as government mandates propelled the industry-wide adoption.

The US General Administration (GSA) has continuously published BIM guides for the industry, including a 3D-4D-BIM overview, spatial program validation, 3D laser scanning, 4D phasing, energy performance and operations, circulation and security validation, building elements, and facility management. Other initiatives and bodies include the National Institute of Building Services (NIBS), which interfaces between the government and the non-government sector. The NIBS, in an alliance with building SMART, developed the NBIMS-US (National BIM Standard-United States).

3.4.4 Indonesia

Indonesia has made several efforts to ensure a macro-level BIM adoption. However, it has been primarily driven by the government. As articulated by Sopaheluwakan and Adi (2020), the formal introduction was done in 2017 by the Ministry of Public Works and Public Housing (PUPR), Indonesia. As part of the efforts to achieve macro-level implementation, the ministry developed a BIM roadmap, formulated BIM steering teams, and mandated BIM usage on government projects (BIM PUPR, 2018). The BIM steering team was saddled with three objectives as it relates to BIM implementation in Indonesia, namely, to formulate roadmaps and implementation strategies, to prepare guidelines and policy, and to create awareness and sensitisation through BIM socialisation and workshops.

3.4.5 Scandinavian Region

This region has been referred to as BIM leaders by several studies. Specifically, Norway, Denmark, and Finland were early adopters of BIM. The government has majorly driven BIM implementation in this region. In the Finnish and Danish construction industry, huge investment by the government and the formulation of policies and guidelines and mandates have impacted greatly on BIM implementation (Al-Btoush & Btoosh, 2020; Smith, 2014).

3.4.6 Asian Region

Ismail et al. (2017) conducted a study on BIM implementation in Asian developing countries. The study findings revealed a not-too-encouraging BIM implementation in the selected countries. However, there are other countries with good BIM implementation. China, Japan, and Singapore are reported to have good BIM implementation nationally. According to Smith (2014), Singapore has dedicated bodies that developed BIM implementation strategies, and there is also a huge investment by the government and government mandate to achieve BIM implementation. Japan is also advanced in terms of BIM adoption, and contractors in the Japanese construction industry observed a positive return on investment. Although the study classified China to be in its early stages of BIM adoption, there is a BIM steering body and the development of BIM standards, among others, to achieve macro-level BIM implementation.

3.5   Summary

This chapter reviewed BIM implementation strategies from a global perspective. This is because most developing countries are not part of the BIM leading countries, and some do not have sufficient information on their BIM effort and strategies, among others. Unlike developed economies and BIM-leading countries, developing countries are still facing several challenges preventing them from achieving macro-level BIM implementation. From this chapter, it is also critical to note that government involvement is critical to achieving BIM implementation. Furthermore, the development of BIM roadmaps and contextually developed BIM policies and standards are very critical to achieving BIM implementation. Another step in achieving BIM implementation is the formulation of BIM steering committees to drive the implementation efforts.

References

Adekunle, S. A., Aigbavboa, C., Ejohwomu, O., Ikuabe, M., & Ogunbayo, B. (2022). A Critical Review of Maturity Model Development in the Digitisation Era. *Buildings 2022, Vol. 12, Page 858*, *12*(6), 858. https://doi.org/10.3390/BUILDINGS12060858

Adekunle, S. A., Aigbavboa, C., & Ejohwomu, O. (2020, December). BIM Implementation: Articulating the Hurdles in Developing Countries. In *Proceedings of the 8th International Conference on Innovative Production and Construction (IPC 2020): The Hong Kong University of Science and Technology*, 47–54.

Al-Btoush, M. A. K., & Btoosh, J. A. A. Al. (2020). BIM Adoption Strategies – The Case of Jordan. *International Journal of Civil Engineering and Technology (IJCIET)*, *10*(07), 343–348.

Autodesk, ConsultAustralia, & AIA. (2010). BIM in Australia. In *Report on BIM-IDP forums* (Issue November).

BIM PUPR. (2018). *PANDUAN Adopsi BIM dalam Organisasi*.

British Standards Institution (BSI). (2013). PAS 1192-2: 2013 Specification for Information Management for the Capital / Delivery Phase of Construction Projects Using Building Accept. In *British Standards Institution (BSI)*.

Edirisinghe, R., & London, K. (2015). Comparative Analysis of International and National Level BIM Standardization Efforts and BIM adoption. *Proc. of the 32nd CIB W78 Conference 2015, 27th–29th October 2015, Eindhoven, The Netherlands, June 2016*, 149–158.

Ismail, N. A. A., Chiozzi, M., & Drogemuller, R. (2017). An Overview of BIM Uptake in Asian Developing Countries. *AIP Conference Proceedings 1903*. https://doi.org/10.1063/1.5011596

Jung, W., & Lee, G. (2015). The Status of BIM Adoption on Six Continents. *International Journal of Civil and Environmental Engineering*, *9*(5), 512–516. https://www.semanticscholar.org/paper/The-Status-of-BIM-Adoption-on-Six-Continents-Jung-Lee/ea0d7a32ebe25d64509e4224e6be9371c6aa1369#paper-header

Khosrowshahi, F., & Arayici, Y. (2012). Roadmap for Implementation of BIM in the UK Construction Industry. *Engineering, Construction and Architectural Management*, *19*(6), 610–635. https://doi.org/10.1108/09699981211277531

Kubba, S. (2017). Building Information Modeling (BIM). In *Handbook of Green Building Design and Construction*. https://doi.org/10.1016/B978-0-12-810433-0.00005-8

McGraw Hill. (2012). *The Business Value of BIM in North America: Multi-Year Trend Analysis and User Ratings (2007-2012)*. www.construction.com

McGraw Hill. (2014). *The Business Value of BIM for Construction in Major Global Markets: How Contractors Around the World Are Driving Innovation With Building Information Modeling*. www.construction.com

Smith, P. (2014). BIM Implementation - Global Strategies. *Procedia Engineering*, *85*, 482–492. https://doi.org/10.1016/J.PROENG.2014.10.575

Sopaheluwakan, M. P., & Adi, J. W. (2020). Adoption and Implementation of Building Information Modeling (BIM) by the Government in the Indonesian Construction Industry. *IOP Conf. Series: Materials Science and Engineering*. https://doi.org/10.1088/1757-899X/930/1/012020

Succar, B. (2010). The Five Components of BIM Performance Measurement. *Proceedings of CIB World Congress, Salford*. https://doi.org/10.1136/bmj.3.5560.312-a

Zaini, N., Zaini, A. A., Tamjehi, S. D., Razali, A. W., & Gui, H. C. (2019). Implementation of Building Information Modeling (BIM) in Sarawak Construction Industry: A Review. *IOP Conference Series: Earth and Environmental Science*. https://doi.org/10.1088/1755-1315/498/1/012091

4 Overview of BIM as a Driver for Optimum Performance of the Construction Industry

4.1 Introduction

Despite the inherent benefits and the widespread adoption rate in developed countries of building information modelling (BIM), it is the most common yardstick for executing buildings throughout their life cycle (Bryde et al., 2013). However, the definition of this technology is varied. Different researchers have described it from diverse perspectives. Hence, there exists no common and generally accepted definition. This has been observed to present a problem in measuring BIM effectiveness and prohibit collaboration within stakeholders (Barlish & Sullivan, 2012). Table 4.1 provides some literature attempts to describe BIM; it is to be noted that

Table 4.1 Definition of BIM

Definition	Source
BIM is a collaborative tool that cuts through all stakeholders in the construction industry.	Eastman (2011)
BIM is the use of a shared digital representation of a built asset (built assets include but are not limited to buildings, bridges, roads, and process plants) to facilitate design, construction, and operation processes to form a reliable basis for decisions.	ISO 19650-1:2018
BIM is a system approach to design that creates dependencies and allows for better collaboration and innovative ways of working between designers and engineers.	Chimhundu (2015)
BIM is a data-rich, intelligent, and object-oriented parametric building modelling tool.	Gao et al. (2019)
"A Building Information Model (BIM) is a digital representation of physical and functional characteristics of a facility. As such, it serves as a shared knowledge resource for information about a facility forming a reliable basis for decisions during its life-cycle from inception onward. The BIM is a shared digital representation founded on open standards for interoperability."	BuildingSMART alliance (2010)
BIM is a digital form of construction and asset operations. It brings together technology, process improvements, and digital information to radically improve client and project outcomes and asset operations.	EUBIM (2017)

(Continued)

DOI: 10.1201/9781003373919-6

Table 4.1 (Continued)

Definition	Source
Building information modelling (BIM) is an intelligent 3D model-based process that gives architecture, engineering, and construction (AEC) professionals the insight and tools to plan, design, construct, and manage buildings and infrastructure more efficiently.	Autodesk
Building information modelling (BIM) is a process for creating and managing information on a construction project across the project life cycle.	NBS (2016)
Building information modelling (BIM) is a set of interacting policies, processes, and technologies generating a "methodology to manage the essential building design and project data in digital format throughout the building's life-cycle."	Bhuskade (2015)
BIM is a very broad term that describes the process of creating digital information about a building or asset (such as a bridge, highway, tunnel, and so on).	Dakhil et al. (2015)
BIM is a product of the general improvement in technology created to encourage teamwork and collaboration during the design and construction period.	Onungwa and Uduma-Olugu (2017)

this is not exhaustive, and it covers the range of years 2010–2020. This captures recent definitions that must have been refined in line with the developments experienced in the BIM technological space.

A careful study of the various definitions shows that various authors view BIM from different perspectives. Generally, they see BIM as a platform, technology, tool, system, and process. This can allude to the fact that the authors view BIM from different perspectives, including manufacturer, contractor, designer, etc. BIM is perceived differently by different stakeholders in the construction industry. This is dependent on the application of BIM in the respective fields. Five capabilities essentially characterise BIM: a clear representation of design; documentation and information management; inbuilt intelligence, analysis, and simulation tool; and collaboration and integration (Gu et al., 2014).

For this study to achieve an encompassing definition, the study examined the meaning of a definition. The definition generally must have three parts: the term being defined, its classification, and its characteristics or functions. Thus, BIM can be defined as a technological tool that provides a collaborative platform among construction project stakeholders to achieve a seamless, synchronised, and real-time execution of processes and information exchange throughout a project life cycle.

4.2   History of BIM

The first interaction between man and the computer in the construction industry happened in 1963 when the first computer-aided design (CAD) was developed. According to Sutherland (2003), it was the breakthrough in human–computer interaction. Ever since, several attempts have been made to foster human–computer

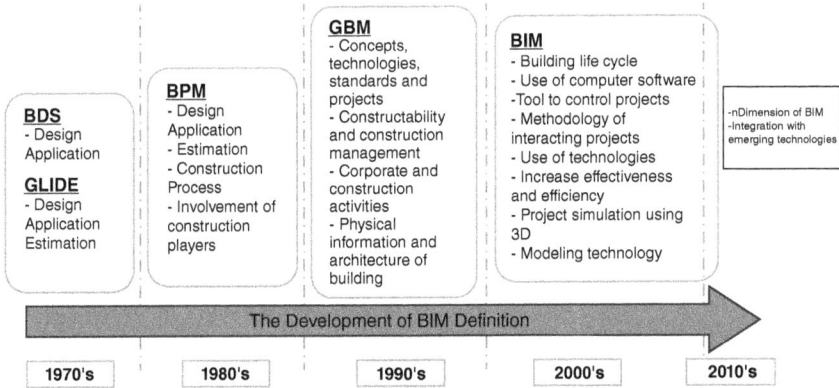

Figure 4.1 The development of BIM definition (Adapted from Aryani et al., 2014)

communication in the construction industry. What is known today as the BIM started as an idea by Eastman (1975) in his work on the use of computers in building design. He described the functions of BIM beyond what it has presently accomplished. He described it as the building description system (BDS). However, the first use of the term "building information modelling" appeared in the research publication by Van Nederveen and Tolman (1992). *Figure 4.1* chronicles the trend of BIM definition over the years. The capability of BIM has experienced continued robustness and more accommodating of several functions to produce effectiveness in the construction industry. Its more robust capability helps in overcoming the many complexities inherent in the construction project process.

Although it took 25 years before BIM became a reality after Eastman's work, there have been different adoption strategies and adoption rates worldwide, as stated in earlier chapters. Also, the research outputs in different countries differ. The first research output on BIM was in 2004 (Liu et al., 2019), although Zhao (2017) opines that it was in 2005. Both studies are literature review studies with data from the web of science (WOS) database; however, the search terms and duration range of Liu et al.'s (2019) study are more encompassing than that of Zhao's.

Figure 4.2 extracted from the Scopus database shows that the highest research outputs in BIM research are from the United States. This is followed by China and the UK. However, the most productive African country is Nigeria, followed by South Africa.

4.2.1 BIM and the Fourth Industrial Revolution

With the advent of the fourth industrial revolution (4IR), many technologies have been introduced. These technologies have disrupted many sectors and engineered catalytic changes. This revolution of technology has also hit the construction industry, which is peculiar in terms of its process and products. Among the many uniqueness of the construction industry is its ability to turn around the economy; thus, it was named the cornerstone of the world's economy (Gerbert et al., 2016). Other peculiarities are observed in its busy nature and fragmented nature, among others.

Documents by country or territory
Compare the document counts for up to 15 countries/territories.

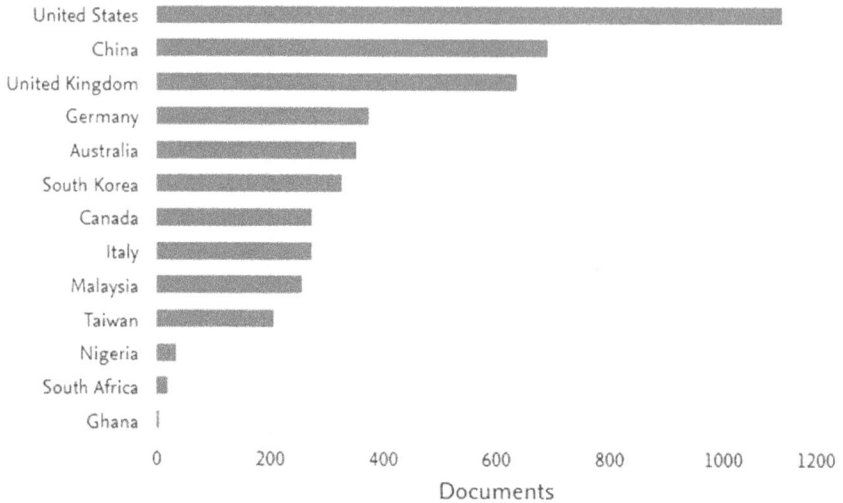

Figure 4.2 Country-wise research outputs
Source: Scopus.

The construction industry is busy because it involves executing many activities by different parties – skilled and unskilled. These activities are performed with the intent to actualise the project objectives. As a result, they occur during the preconstruction stage (design, procurement, and contractor selection, among others; generally, the planning stage) and construction stage (inception of physical construction to physical completion and handover) post-construction phase of the project. The smooth execution of these activities requires proper coordination to achieve the synergy of tasks and functions (Darshi de Saram & Ahmed, 2001).

The effective working of the construction process requires coordination and synergy to avert conflict. Consequently, the collaboration by parties on a project is an important ingredient to transmit the client's dream to reality. Ineffective synergy within project parties results in delay, thus hampering the progress of the project (Adekunle et al., 2020a). Cost and time overrun are a resultant effect of the failure of project parties to work together effectively. This is a harbinger of friction and an unhappy project team. Most times, project parties' problems are caused by human error, untimely execution of work, inaccurate output, and others, which are all due to the traditional method of performing functions.

However, the introduction of the industrial and technological revolution changed the process and products of the construction industry significantly. The technological age provided solutions to many of the erstwhile challenges facing the construction industry. The advantage of technology to the construction industry has been established in the literature. Value creation, accuracy, faster production, less conflict, and error reduction were established advantages of technology on the process in the construction industry (Ingi, 2009). The product in the construction industry has been observed to experience great advancement due to the introduction of technology.

The turnaround of the construction industry is anchored on technology as experienced in other industries, for instance, manufacturing. Thus, the construction industry cannot afford to neglect the adoption of technology in overcoming its challenges because technology will turn around the industry and provide efficiency and quality (Castagnino et al., 2016; Gerbert et al., 2016).

It is becoming extremely difficult to ignore the importance of BIM considering the recent advancement in technology in the construction industry. BIM is constantly in the discourse, albeit from diverse perceptions.

4.3   What BIM Is Not

Many studies have attempted to define what BIM is; however, the aspect of defining what it is not has received little attention. This section is not dwelling on the false belief or try to debunk such. However, it is concerned with the functions BIM cannot perform. The following are some of the assertions of what BIM is not:

i BIM is not a silver bullet that solves the entire problem of the construction industry. Stakeholders in the construction industry are still required to perform their duties. They are to be more meticulous with data handling as the smooth running and accuracy of the technology depend on their input. There are still risks inherent to the adoption of BIM by parties.
ii There is a difference between BIM and CAD. BIM was not meant to be CAD (Zyskowski, 2009).
iii The BIM platform is a collaborative information exchange platform. It is likened to the paper sharing era when project drawing constitutes the project document, but not the contract. BIM produces the platform for information sharing but not a contract document (Ashcraft, 2008). BIM platform does not provide the space for the signing of contracts and parties' obligations.

It is worthy of note that point (iii) above can be achieved through integration with other emergent technologies. This has become possible in the 4IR era.

4.4   BIM Adoption: Benefits, Drivers, and Barriers

The diffusion of technological innovations is well captured by Rogers (1962) in his diffusion of innovation theory. Diffusion, according to Rogers, is the process through which an innovation is communicated among actors of a social system and via channels. This theory seeks to explain how innovation diffuses and the rate at which it does. Diffusion deals with the spread of innovation among the actors in an ecosystem. The theory categorises innovation adoption by actors into four categories: early adopters, early majority, late majority, and laggards.

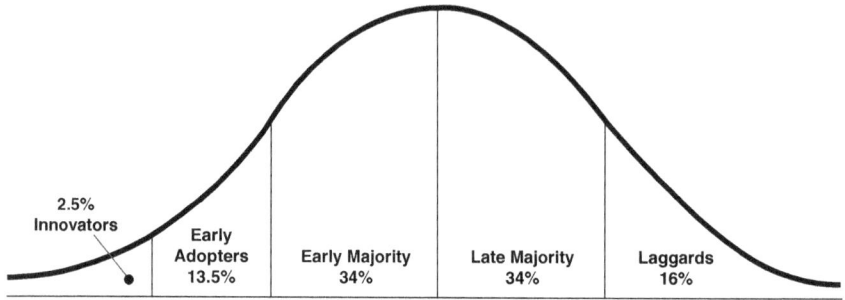

Figure 4.3 Diffusion of innovation (Rogers, 1962)

The theory posits that for an innovation to gain widespread in an ecosystem, there are four elements: innovation, communication channel, time, and social system. Time is one of the four elements required for innovation diffusion to be achieved. However, central to achieving the diffusion is the decision by actors; their decision is dependent on the knowledge they possess of the innovation. Thus, moving from one category of the adopter to another is dependent on the knowledge of the inherent benefits of the innovation (Figure 4.3).

The ability to successfully convince and change his perception at the knowledge stage of the innovation decision-making process is hinged on the benefits inherent in the innovation.

4.5   Benefits of BIM

The task-technology fit theory posits that technology adoption is more likely if it possesses capabilities that match the tasks that the actor wants to perform (Goodhue & Thompson, 1995). Although the theory termed the yardstick as "positive impact" and not adoption, an innovation cannot impact until utilised.

By extension, BIM adoption is also hinged on its inherent ability to match the tasks to be performed in the construction industry. The many benefits of BIM to the construction industry have been studied in the literature, and these identified benefits cut across every phase of the project life cycle. In its classification of the outline plan of work in the light of BIM, the Royal Institute of British Architects (RIBA) classified the plan of work and tasks as follows in Table 4.2. BIM can also be employed in building renovation projects (Joblot et al., 2019) and building sustainability assessments (Carvalho et al., 2019). BIM is assessed in terms of the benefits from the tasks outlined in Table 4.2. The benefits are visible, measurable, and tangible. Evidenced benefits of BIM are drivers for its adoption as it changes the perception and persuades actors to adopt it.

The application of BIM in the project's life cycle and other attendant technological tools per stage is illustrated in Figure 4.4.

The BIM benefits transcend professional barriers as it enhances the output and efficiency of the project stakeholders collectively. Generally, BIM adoption is poised to make the construction industry more productive and meet the

Table 4.2 Construction project life cycle tasks (RIBA, 2012)

RIBA work stage			Core BIM tasks
Preparation	A	Appraisal	• Advise client on the purpose of BIM, including benefits and implications.
	B	Design brief	• Agree on the level and extent of BIM, including 4D (time), 5D (cost), and 6D (FM) following software assessment.
			• Advise client on Integrated Team scope of service in totality and for each designer, including requirements for specialists and a BIM Model Manager appointment.
			• Define long-term responsibilities, including ownership of the model.
			• Define BIM inputs and outputs and scope of post-occupancy evaluation (Soft Landings).
			• Identify the scope of and commission of BIM surveys and investigation reports.
			• Data drop 1.
Design	C	Concept	• BIM pre-start meeting.
			• Initial model sharing with Design Team for strategic analysis and options appraisal.
			• BIM data used for environmental performance and area analysis.
			• Identify key model elements (e.g., prefabricated component) and create concept-level parametric objects for all major elements.
			• Enable design team access to BIM data.
			• Agree on the extent of performance-specified work.
			• Data drop 2.
	D	Design development	• Data sharing and integration for design coordination and detailed analysis, including data links between models.
	E	Technical design	• Integration/development of generic/bespoke design components.
			• BIM data used for environmental performance and area analysis.
			• Data sharing for design coordination, technical analysis, and addition of specification data.
			• Export data for Planning Application.
			• 4D and/or 5D assessment.
			• Data drop 3.
Pre-construction	F	Production information	• Export data for Building Control Analysis.
	G	Tender documentation	• Data sharing for the conclusion of design coordination and detailed analysis with subcontractors.
	H	Tender action	• Detailed modelling, integration, and analysis.
			• Create production-level parametric objects for all major elements (where appropriate and information exists this may be based on tier 2 supplier's information).
			• Embed specification to model.

(*Continued*)

Table 4.2 (Continued)

RIBA work stage			Core BIM tasks
			• Final review and sign-off of the model.
			• Enable access to the BIM model to the contractor(s).
			• Integration of subcontractor performance-specified work model information into BIM model data.
			• Review construction sequencing (4D) with the contractor.
			• Data drop 4.
Construction	J	Mobilisation	• Agree on timing and scope of "Soft Landings."
	K	Construction to practical completion	• Coordinate and release of "End of Construction" BIM record model data.
			• Use of 4D/5D BIM data for contract administration purposes.
			• Data drop 5.
Use	L	Post-practical completion	• FM BIM model data issued as asset changes are made.
R&D	M	Model maintenance and development	• Study of parametric object information contained within BIM model data.
			• Data drop 6

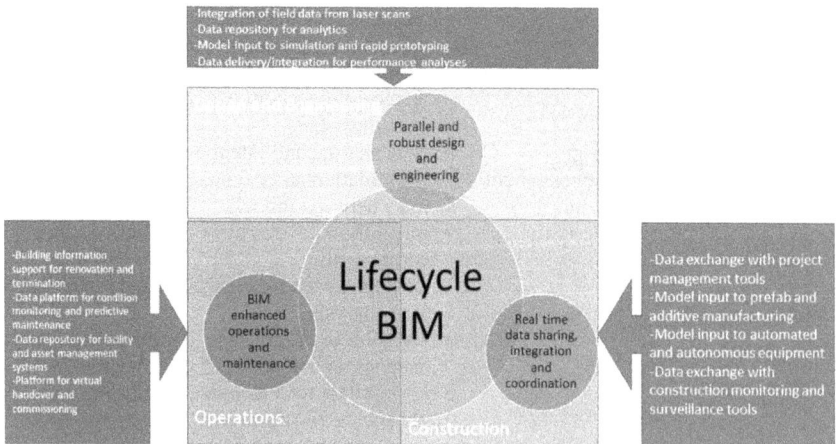

Figure 4.4　BIM application in the construction life cycle (Bühler et al., 2016)

ever-changing client demand for infrastructure. The demand is due to the infra-structure gap (WEF & BCG, 2016).

　　The bedrock of BIM adoption and effectiveness is data exchange at every stage of the project life cycle. Figure 4.4 shows that every stage of the project life cycle depends on the previous stage's data input. Thus, it requires collaboration and ade-quate stakeholder motivation (WEF & BCG, 2018). Figure 4.4 and Tables 4.2 and 4.3 show that BIM possesses the capabilities fit for the tasks in the construction process.

Table 4.3 Benefits of BIM adoption (Adekunle et al., 2022; Aung, 2018; Babatunde et al., 2018; BuildingSMART alliance, 2010; Cassino et al., 2013; Chimhundu, 2015; Eastman et al., 2008; Mostafa et al., 2018; Olsen & Taylor, 2017)

Benefits

Process			People	Overall
Preconstruction (concept, feasibility, and design stage)	Construction and fabrication benefits	Post-construction		
Increased building performance and quality	Synchronise design and construction planning	Better manage and operate facilities	Quantity surveyor in quantity take-off, estimating, automated BoQ preparation	Providing clients with better value for money, gaining clients confidence, increased productivity, reduced waste on projects
Earlier and more accurate visualisations of a design	Reduction or error through design management and clash detection.	Integrate with facility operation and management systems	Increased profit and enhanced image for contractors	Reduction of rework, construction cost, project duration, and reduced safety risks
Automatic low-level corrections when changes are made to design	React quickly to design or site problems	Better scheduled maintenance and easy access to information	Better focus on workers supervising	Evaluating proposed construction methodologies in terms of practicality and ease of construction
Generate accurate and consistent 2D drawings at any stage of the design	Use design model as a basis for fabricated components			Improved communication within project delivery teams (professional consultants and contractors) and effective project team collaboration.
Earlier collaboration of multiple design disciplines	Better implementation and lean construction techniques			Better storage of design data throughout the entire built asset's life cycle
Easily check against the design intent	Synchronise procurement with design and construction			Ease in outlining project material and resource requirements
Extract cost estimates during the design stage				Better planning of projects before construction on site.
Improve energy efficiency and sustainability				Providing price certainty for construction projects. Promoting the competitiveness of construction activities and growth.

4.6   Barriers to BIM Adoption

Despite the inherent and known benefits of BIM, it has not enjoyed widespread adoption. Although this is not peculiar to BIM, this is a peculiar trend with innovation adoption in the construction industry. Compared to other industries, the construction industry is slow in adopting new innovations (Bühler et al., 2016). The study posited that the construction had been the same for over 50 years, and its failure to adopt innovations will make it impossible to match up with clients' infrastructural demand.

Similarly, the adoption of BIM has experienced some barriers in the construction industry. Various studies have been conducted in this regard, and they identified diverse barriers to the BIM adoption drive (Table 4.4).

The barriers to implementation can be classified into process, technological, cultural, and financial barriers. The success of BIM adoption depends on the ability to overcome them at both macro and micro levels.

To ascertain if the barriers are the same irrespective of geographical location, this study compared identified BIM barriers in the literature.

Table 4.5 presents the barriers to BIM adoption globally in selected countries. The countries selected cut different continents. However, it has been established that there are diverse perspectives regarding BIM barriers between BIM users and non-users. This distinction was observed by Eadie et al. (2014) while analysing BIM barriers among organisations in the United Kingdom. Extending this distinction in perspective to countries, Jung and Lee (2015) opined that there exist different levels of BIM adoption globally. However, a comparison of barriers globally has not been conducted to ascertain if there exists a difference in barriers.

Table 4.4 The identified barrier to BIM implementation in the literature

Source	Barrier
Elagiry et al. (2019)	Inaccurate data collection and difficulties in collaboration
Armah (2015)	Set-up cost, lack of awareness, risk exposure, and culture resistance
McGraw Hill (2014)	Structural barriers
Eastman et al. (2008)	High training cost, lopsided adoption among stakeholders, legal barriers, lack of trained personnel in HEIs, and interoperability issues
NBS (2012)	The proliferation of knowledge and confusion about what BIM is.
Mandhar and Mandhar (2013)	Culture, scepticism, and obsolete school curriculum.
Chimhundu (2015)	Reluctance to change and IT infrastructural limitations.
Ayinla and Adamu (2018)	Lack of client interest.
Memon et al. (2014)	Data ownership.
Oraee et al. (2019)	Dynamics and fragmented nature of the construction industry, lack of contractual standards around BIM models, the difference in organisational structure, and unequal understanding of collaboration within the industry.

Table 4.5 Geographical BIM barriers (Adekunle et al., 2020; Amuda-Yusuf et al., 2017; Fadason et al., 2018; Khosrowshahi & Arayici, 2012; Memon et al., 2014; Mutonyi et al., 2018; Rogers et al., 2015; Sawhney, 2014; Stanley & Thurnell, 2014)

	United Kingdom	Nigeria	India	Kenya	Malaysia	New Zealand
Barriers						
Unfamiliarity with BIM use	✓					
Intangible benefits to warrant use	✓					
Cultural resistance	✓		✓	✓		
Not offering enough benefits to warrant the use	✓					
Reluctance to new workflows	✓		✓		✓	
Lack of capital	✓	✓	✓	✓	✓	✓
Cost outweighs benefits	✓					
No demand for BIM use	✓	✓			✓	
Lack of BIM education/training		✓				
Lack of government support		✓	✓			
Lack of standards		✓		✓		✓
Lack of infrastructure		✓				
Lack of experts		✓	✓		✓	
Initial effort and time required			✓		✓	
Lack of awareness		✓		✓	✓	
Interoperability				✓		✓
Legal related issues		✓			✓	
Lack of communication		✓			✓	
Insufficient professional guidance					✓	

Table 4.5, however, shows that lack of capital is the common barrier to BIM adoption globally. However, aside from this common barrier, it is evident that barriers are environment-specific; this supports the work of Eadie et al. that perceptions vary between users and non-users. For instance, lack of standards and lack of awareness are peculiar to the African continent.

4.7   BIM Ecosystem

The advent of the BIM required the interaction of technology, human actors, and processes. Gu et al. (2014) stated that there must be coevolution of people, processes, and products to achieve a balanced BIM ecosystem. These different dimensions are interdependent and in constant interaction. This interaction is both intra- and inter-dimensional.

The BIM process involves many activities; thus, many tools are required. These BIM technologies include design authoring, data management, visualisation, parametric modelling, data capturing, simulation, scheduling, and data management. These technologies making up the BIM process require different tools. This requires that different software from different manufacturers are employed and used together.

Tables 4.6 and 4.7 show different BIM tools and platforms in the BIM ecosystem. From the array of tools and platforms, it is evident that numerous tools are available to stakeholders. Although this is good, it can also be a daunting task for stakeholders to choose. Stakeholders will be required to consider many factors before making their choice.

Table 4.6 BIM platforms (Eastman, 2011; Mbarga & Mpele, 2019)

Platforms	Specific tools	File collaboration formats
ArchiCAD (Graphisoft)	ArchiCAD	IFC, BCF, OBDC, DWF, NWC, SMC, 3DS, 3DM, SKP, KML, OBJ, STL, ...
AutoCAD (Autodesk)	AutoCAD Architecture, AutoCAD MEP, AutoCAD Electrical, AutoCAD Civil 3D, AutoCAD P&D et Plant 3D.	DGN, DWG, DWF, DXF, IFC, ...
Bentley (Bentley)	Bentley Architecture, Bentley PowerCivil, RAM Structural System, ...; GEOPAK Civil Engineering Suite, Bentley Building Electrical Systems, Facility Information Management, ...; Bentley view	IFC, CIS/2, STEP, DWG, DXF, U3D, 3DS, Rhino 3DM, IGES, SAT, STEP AP203/AP214, STL, OBJ, KML, SKP, ...
Cype (Cype)	CYPECAD, CYPETHERM, CYPEPROJECT	IFC, CIS/2, DXF, DWG, ...
Revit (Autodesk)	Revit Architecture, Revit Structure, Revit MEP, Naviswork	IFC, gbXML, RVA, DWG, DWF, DGN, SKP, IES, FBX, ODBC, SAT, ADSK, BIMétré, ...
Tekla (Trimble)	Tekla Structures, Plancal Nova, Tekla BIMsight	DWG, DXF, CIS/2, STP, XML, IFC, IGES, DGN, ODBC, SAP, SDNF, SDF, STEP,...
Vectorworks (Nemetschek)	Architect, Designer, Landmark, Spotlight, Machine design, Solibri Model Viewer, Solibri Model Checker	IFC, DXF/DWG, STL, 3DS, Revit, SKP, ...

Table 4.7 Some BIM tools (Eastman, 2011; Mbarga & Mpele, 2019)

S/N	Tools	BIM tools
1	Tools for rebuilding BIM models from existing	Tripod (Measurix), Viz"All (All Systems)
2	Tools for preliminary design	FreeCAD, Rhinoceros (Robert McNeel & Associates), SketchUp (Trimble), SolidWorks Premium (Dassault Systemes)
3	Tools for architectural design	Allplan Architecture (Allplan/ Nemetschek), ArchiCAD (Graphisoft/ Nemetschek), Bentley Architecture (Bentley), Revit Architecture (Autodesk), Vectorworks Architect (Vectorwork/ Nemetschek)
4	Tools for structural modelling and analysis	Allplan Engineering (Allplan/ Nemetschek), CYPE 3D (Cype), CYPECAD (Cype), Revit Structure (Autodesk), Robot Structural Analysis (Autodesk), Scia Engineer (Scia/ Nemetschek), STAAD-Pro (Bentley), Tekla structure (Tekla/ Trimble)

(Continued)

Table 4.7 (Continued)

S/N	Tools	BIM tools
5	Tools for mechanical, electrical, and plumbing (MEP)	Bentley Hevacomp Mechanical Designer (Bentley), CYPETHERM (Cype), Revit MEP (Autodesk), DDS-CAD MEP (Nemetschek)
6	Tools for model review and coordination	Bentley view (Bentley), Naviswork (Autodesk), Solibri Model Checker (Nemetschek), Tekla BIMsight (Trimble)
7	Tools for cost estimation	WinQUANT Q4 (Attic+), Glodon Takeoff for Architecture and Structure (Glodon Software Company Limited), CostOS Estimating (Nomitech)
8	Tools for thermal analysis	ArchiWIZARD (Graitec), Bentley Hevacomp Mechanical Designer (Bentley), ClimaWin (BBS Slama)
9	Tools for environmental impact analysis	Elody-eveBIM (CSTB), IDA ICE (EQUA Simulation AB), Energy Plus
10	Tools for facility management	ACTIVe3D Facility Server (Sopra Steria), Allfa Web (Allplan), ArchiFM (Graphisoft)

4.8   Summary

BIM as a critical factor in the transformation of the construction industry in the digital age was focused on in this chapter. The chapter provided a broad overview of BIM. It explores the history of BIM and the trend in information management metamorphosing into BIM. Furthermore, the various definitions of BM in literature were explored. Ultimately, the book formulated a definition of BIM in the context of this publication. Also, the various BIM benefits to the construction industry were identified, and the BIM barriers confronting the construction industry were identified. This chapter also explored the BIM ecosystem and identified some of the tools and standards; however, it should be noted that the list is not exhaustive as the BIM ecosystem is constantly changing and dynamic based on the advancement of technology. The chapter provides insight into BIM and its importance to the construction industry.

References

Adekunle, S., Aigbavboa, C., Akinradewo, O., Ikuabe, M., & Adeniyi, A. (2022). A Principal Component Analysis of Organisational BIM Implementation. *Modular and Offsite Construction (MOC) Summit Proceedings*, 161–168. https://doi.org/10.29173/MOCS278

Adekunle, S. A., Aigbavboa, C., & Ejohwomu, O. A. (2020a). Improving Construction Project Performance in Developing Countries: Contractor Approach. *3rd European and Mediterranean Structural Engineering and Construction Conference 2020, Euro-Med-Sec 2020*, 7(1). https://doi.org/10.14455/isec.res.2020.7(1).con-14

Adekunle, S. A., Aigbavboa, C., & Ejohwomu, O. (2020b, December). BIM Implementation: Articulating the Hurdles in Developing Countries. In *Proceedings of the 8th International Conference on Innovative Production and Construction (IPC 2020)*, The Hong Kong University of Science and Technology, 47–54.

Amuda-Yusuf, G., Adebiyi, R. T., Olowa, T. O. O., & Oladapo, I. B. (2017). Barriers to Building Information Modelling Adoption in Nigeria. *Journal of Research Information in Civil Engineering, 14*(2), 1574–1591.

Armah, N. N. O. (2015). *Assessing the Benefits and Barriers of the Use of Building Information Modelling (BIM) in the Ghanaian Building Construction Industry.* Kwame Nkrumah University of Science and Technology, Kumasi.

Aryani, A. L., Brahim, J., & Fathi, M. S. (2014). The Development of Building Information Modeling (BIM) Definition. *Applied Mechanics and Materials, 567*(August), 625–630. https://doi.org/10.4028/www.scientific.net/AMM.567.625

Ashcraft, H. W. (2008). Building Information Modeling: A Framework for Collaboration Project Management. *Construction Lawyer, 28*(3). https://www.hansonbridgett.com/-/media/Files/Publications/bim_building_information_modeling_a_framework_for_collaboration.pdf

Aung, Y. (2018). *Practices and Impacts of Building Information Modelling (BIM) for Construction Project in Singapore.* Birmingham City University.

Ayinla, K. O., & Adamu, Z. (2018). Bridging the Digital Divide Gap in BIM technology Adoption. *Engineering, Construction and Architectural Management, 25*(10), 1398–1416. https://doi.org/10.1108/ECAM-05-2017-0091

Babatunde, S., Ekundayo, D., Babalola, O., & Jimoh, J. (2018). Analysis of the Drivers and Benefits of BIM Incorporation into Quantity Surveying Profession: Academia and Students' Perspectives. *Journal of Engineering, Design and Technology.* https://doi.org/10.1108/JEDT0420180058

Barlish, K., & Sullivan, K. (2012). How to Measure the Benefits of BIM — A Case Study Approach. *Automation in Construction, 24*, 149–159. https://doi.org/10.1016/j.autcon.2012.02.008

Bhuskade, S. (2015). Building Information Modeling (BIM). *International Research Journal of Engineering and Technology.* www.irjet.net

Bryde, D., Broquetas, M., & Volm, J. M. (2013). The Project Benefits of Building Information Modelling (BIM). *International Journal of Project Management, 31*(7), 971–980. https://doi.org/10.1016/j.ijproman.2012.12.001

Bühler, M. M., Almeida, P. R. de, Solas, M. Z., Beck, J. M., & Renz, A. (2016). *Industry Agenda Shaping the Future of Construction A Breakthrough in Mindset and Technology Prepared in collaboration with The Boston Consulting Group.* https://doi.org/https://doi.org/10.13140/RG.2.2.32674.96964

BuildingSMART alliance. (2010). *Building Information Modeling Execution Planning Guide.* http://www.buildingsmartalliance.org/nbims.

Carvalho, J. P., Bragança, L., & Mateus, R. (2019). Optimising Building Sustainability Assessment Using BIM. *Automation in Construction, 102*(March), 170–182. https://doi.org/10.1016/j.autcon.2019.02.021

Cassino, K. E., Bernstein, H. M., Asce, F., Ap, L., Russo, M. A., Advisor, A. E., Jones, S. A., Laquidara-Carr, D., Manager, W. T., Operations, C., Ramos, J., Lorenz, A., Washington, M., Yamada, T., Fitch, E., Associate, G., Communications, B., Knapschaefer, J., & Barnett, S. (2013). *SmartMarket Report McGraw Hill Construction Information Mobility: Improving Team Collaboration Through the Movement of Project Information SmartMarket Report Executive Editor.* www.construction.com

Castagnino, S., Rothballer, C., & Gerbert, P. (2016). What's the Future of the Construction Industry? *World Economic Forum.* https://www.weforum.org/agenda/2016/04/building-in-the-fourth-industrial-revolution/

Chimhundu, S. (2015). *A Study on the BIM Adoption Readiness and Possible Mandatory Initiatives for Successful Implementation in South Africa*. University of the Witwatersrand, Johannesburg.

Dakhil, A., Alshawi, M., & Underwood, J. (2015). BIM Client Maturity: Literature Review. *12th International Post-Graduate Research Conference 2015 Proceedings*, June, 229–238.

Darshi de Saram, B. D., & Ahmed, S. M. (2001). *Construction Coordination Activities: What Is Important and What Consumes Time. Journal of Management in Engineering*, *17*(4), 203–213.

Eadie, R., Odeyinka, H., Browne, M., Mckeown, C., & Yohanis, M. (2014). Building Information Modelling Adoption: An Analysis of the Barriers to Implementation. *Journal of Engineering and Architecture*, *2*(1), 77–101. https://doi.org/10.1007/s13398-014-0173-7.2

Eastman, C. (1975). The Use of Computers Instead of Drawings in Building Design. *AIA Journal, January 1975*.

Eastman, C. M. (2011). *BIM Handbook : A Guide to Building Information Modeling for Owners, Managers, Designers, Engineers and Contractors*. Wiley. https://books.google.co.za/books?hl=en&lr=&id=aCi7Ozwkoj0C&oi=fnd&pg=PP7&ots=ZbDeOUz7Iq&sig=yvlZ_9-KVeVMEAmjXEog94FzI7Q&redir_esc=y#v=onepage&q&f=false

Eastman, C, Teicholz, P., Sacks, R., & Liston, K. (2008). *BIM Handbook A Guide to Building Information Modeling for Owners, Managers, Designers, Engineers, and Contractors*. John Wiley & Sons, Inc. https://doi.org/10.1093/nq/s7-II.32.110-e

Elagiry, M., Lasarte, E., & Messervey, T. (2019). BIM4Ren: Barriers to BIM Implementation in Renovation Processes in the Italian Market. *Buildings*, *20*(1), 24. https://doi.org/10.3390/proceedings2019020024

EUBIM. (2017). *Handbook for the Introduction of Building Information Modelling by the European Public Sector Strategic Action for Construction Sector Performance: Driving Value, Innovation and Growth*. www.eubim.eu

Fadason, R. T., Danladi, C. Z., & Akut, K. L. (2018). Challenges of Building Information Modeling Implementation in Africa : A Case Study of the Nigerian Construction Industry. *FIG Congress 2018 Embracing Our Smart World Where the Continents Connect: Enhancing the Geospatial Maturity of Societies Istanbul, Turkey, May*.

Gao, H., Koch, C., & Wu, Y. (2019). Building Information Modelling Based Building Energy Modelling: A Review. *Applied Energy*, *238*(March), 320–343. https://doi.org/10.1016/j.apenergy.2019.01.032

Gerbert, P., Castagnino, S., Rothballer, C., Renz, A., & Filitz, R. (2016). *The Transformative Power of Building Information Modeling*. https://www.bcg.com/publications/2016/engineered-products-infrastructure-digital-transformative-power-building-information-modeling.aspx

Goodhue, D. L., & Thompson, R. L. (1995). Task-Technology Fit and Individual Performance. *MIS Quarterly*, 213–236.

Gu, N., Singh, V., & London, K. (2014). BIM Ecosystem : The Coevolution of Products, Processes, and People. In K. M. Kensek & D. Noble (Eds.), *Building Information Modeling: BIM in Current and Future Practice* (Issue 1, pp. 197–210). John Wiley & Sons, Inc. https://doi.org/10.1002/9781119174752.ch15

Ingi, E. (2009). *Implementation of BIM Danish Experience from Icelandic Perspective*. Technical University of Denmark.

ISO 19650-1:2018. (2018). *Organization and Digitization of Information About Buildings and Civil Engineering Works, Including Building Information Modelling (BIM) - Information Management Using Building Information Modelling* (19650–1). ISO.

Joblot, L., Paviot, T., Deneux, D., & Lamouri, S. (2019). Building Information Maturity Model specific to the renovation sector. *Automation in Construction, 101*, 140–159. https://doi.org/10.1016/j.autcon.2019.01.019

Jung, W., & Lee, G. (2015). The Status of BIM Adoption on Six Continents. *International Journal of Civil and Environmental Engineering, 9*(5), 512–516. https://www.semantic-scholar.org/paper/The-Status-of-BIM-Adoption-on-Six-Continents-Jung-Lee/ea0d7a32e be25d64509e4224e6be9371c6aa1369#paper-header

Khosrowshahi, F., & Arayici, Y. (2012). Roadmap for Implementation of BIM in the UK construction industry. *Engineering, Construction and Architectural Management, 19*(6), 610–635. https://doi.org/10.1108/09699981211277531

Liu, Z., Lu, Y., & Peh, L. C. (2019). A Review and Scientometric Analysis of Global Building Information Modeling (BIM) Research in the Architecture, Engineering and Construction (AEC) Industry. *Buildings, 9*(10), 210. https://doi.org/10.3390/buildings9100210

Mandhar, M., & Mandhar, M. (2013). BIMing the Architectural Curricula – Integrating Building Information Modelling (BIM) in Architectural Education. In *University of Lincoln, United Kingdom (UK)* (Vol. 147, pp. 11–40).

Mbarga, R. O., & Mpele, M. (2019). BIM Review in AEC Industry and Lessons for Sub-Saharan Africa: Case of Cameroon. *International Journal of Civil Engineering and Technology (IJCIET), 10*(5), 930–942. http://www.iaeme.com/IJCIET/index.asp930http://www.iaeme.com/ijciet/issues.asp?JType=IJCIET&VType=10&IType=5http://www.iaeme.com/IJCIET/issues.asp?JType=IJCIET&VType=10&IType=5

McGraw Hill. (2014). *The Business Value of BIM for Construction in Major Global Markets: How Contractors Around the World Are Driving Innovation With Building Information Modeling.* www.construction.com

Memon, A. H., Rahman, I. A., Tun, U., Onn, H., & Memon, I. (2014). *BIM in Malaysian Construction Industry : Status, Advantages, Barriers and Strategies to Enhance the Implementation Level. August.* https://doi.org/10.19026/rjaset.8.1012

Mostafa, S., Kim, K. P., Tam, V. W. Y., & Rahnamayiezekavat, P. (2018). Exploring the Status, Benefits, Barriers and Opportunities of Using BIM for Advancing Prefabrication Practice. *International Journal of Construction Management, 20*(2), 146–156. https://doi.org/10.1080/15623599.2018.1484555

Mutonyi, M., Nasila, M., & Cloete, C. (2018). Adoption of Building Information Modelling in the Construction Industry in Kenya. *Acta Structilia, 25*(2), 1–38. https://doi.org/10.18820/24150487/as25i2.1

NBS. (2012). *National BIM Report 2012 What Will BIM Mean for Design Fees?* 1–20. www.nationalbimlibrary.com

NBS. (2016). What Is BIM? *NBS.* https://www.thenbs.com/knowledge/what-is-building-information-modelling-bim

Olsen, D., & Taylor, J. M. (2017). Quantity Take-Off Using Building Information Modeling (BIM), and Its Limiting Factors. *Creative Construction Conference 2017*, 19–22. https://doi.org/10.1016/j.proeng.2017.08.067

Onungwa, I. O., & Uduma-Olugu, N. (2017). Building Information Modelling and Collaboration in the Nigerian Construction Industry. *JCBM, 1*(2), 1–10. http://journals.uct.ac.za/index.php/jcbm

Oraee, M., Hosseini, M. R., Edwards, D. J., Li, H., Papadonikolaki, E., & Cao, D. (2019). Collaboration Barriers in BIM-Based Construction Networks: A Conceptual Model.

International Journal of Project Management, *37*(6), 839–854. https://doi.org/10.1016/J. IJPROMAN.2019.05.004

RIBA. (2012). *BIM Overlay to the RIBA Outline Plan of Work*. www.aubreykurlansky.co.uk

Rogers, E. M. (1962). *Diffusion of Innovations* (Third Edit). The Free press.

Rogers, J., Chong, H.-Y., & Preece, C. (2015). Adoption of Building Information Modelling Technology (BIM) Perspectives from Malaysian Engineering Consulting Services Firms. *Engineering, Construction and Architectural Management*, *22*(4), 424–445. https://doi. org/10.1108/ECAM-05-2014-0067

Sawhney, A. (2014). *State of BIM Adoption and Outlook in India*. http://www.fig.net/ resources/proceedings/fig_proceedings/fig2014/ppt/ss36/ss36_kavanagh_7434.pdf

Stanley, R., & Thurnell, D. (2014). The Benefits of, and Barriers to, Implementation of 5D BIM for Quantity Surveying in New Zealand. *Australasian Journal of Construction Economics and Building*, *14*(1), 105–117.

Sutherland, I. E. (2003). *Sketchpad: A Man-Machine Graphical Communication System*. http://www.cl.cam.ac.uk/

Van Nederveen, G. A., & Tolman, F. P. (1992). Modelling multiple views on buildings. *Automation in Construction*, *1*, 215–224.

WEF, & BCG. (2016). *Shaping the Future of Construction A Breakthrough in Mindset and Technology*. World Economic Forum.

WEF, & BCG. (2018). *An Action Plan to Accelerate Building Information Modeling (BIM) Adoption*. www.weforum.org

Zhao, X. (2017). A Scientometric Review of Global BIM Research: Analysis and Visualization. *Automation in Construction*, *80*, 37–47. https://doi.org/10.1016/j.autcon.2017.04.002

Zyskowski, P. (2009, February 4). The World According to BIM, Part 1. *Cadalyst*. https:// www.cadalyst.com/cad/building-design/the-world-according-bim-part-1-3780

5   Maturity Model

5.1   Introduction

This chapter is focused on the review of theoretical perspectives on the building information modelling (BIM) maturity model. It discusses and reviews maturity model development. It explores the types and justification for maturity model development and the components of the maturity model, thus providing a roadmap and guideline for maturity model development.

5.2   Overview of Maturity Model

A maturity model is a structured collection of elements that describe the characteristics of an effective process (Carnegie Mellon University, 2005); it describes the pathway for organisational improvements (Proença et al., 2016). Maturity models can also be the collation of process maturity levels from the starting point (immaturity) to the highly mature level (Succar, 2010). Maturity models are important as they provide organisations with an instrument to measure their maturity in an endeavour or process. It thus provides a starting point, a systematic growth, a process to maturity, a common language and platform of assessment, a framework for prioritising strategies, and a way to define what improvement means (Carnegie Mellon University, 2005).

Maturity models are also considered flexible in terms of application and provide a uniform basis for measurement (Succar, 2010). Maturity models can thus be applied irrespective of the organisation or project (size, configuration, or type). It can thus be applied also in the construction industry. However, it produces unique projects with similar procedures. Due to the standardised nature of maturity models, maturity models provide a benchmark for comparison irrespective of organisation or project.

Capability levels generally define maturity models. A capability level is a well-defined evolutionary plateau describing the organisation's capability relative to a particular process area (Carnegie Mellon University, 2005). It refers to an organisation's improvement in each process area (Michael Van Sickle, 2012). Capability levels are cumulative. Thus, a level needs to be achieved before going to the next; it is systematic and stratified. Capability is defined to be the basic ability to

DOI: 10.1201/9781003373919-7

perform a task (Succar, 2010). However, Van Steenbergen et al. (2010) defined capability as an ability to achieve a predefined goal that is associated with a particular maturity level.

Meanwhile, the maturity level is a well-defined evolutionary plateau of achieving a matured process (Carnegie Mellon University, 2005; Paulk et al., 1999). It gives an overall process achievement using the model by the organisation (Michael Van Sickle, 2012). Each maturity level provides a necessary foundation for the effective implementation of processes at the next level. Thus, higher-level processes have less chance of success without the discipline provided by lower levels. Maturity, according to Succar (2010), is the degree of excellence in performing required tasks.

5.3   Necessity for Maturity Models

Maturity models are imperative as a solution when there is a system failure. When a system is not productive and efficient, maturity models assist in assessing and improving effectiveness. According to Carnegie Mellon University (2005), a system failure is characterised by the following, as presented in Table 5.1. These characteristics are evident in the construction industry; thus, a maturity model is important to provide a systematic solution for a more proficient construction industry.

From Table 5.1, it is evident that a quality problem exists in the construction industry process. According to the Juran trilogy, this requires improvement to achieve change. Improvement is:

an activity in which every organisation carries out tasks to make incremental improvements, day after day. Daily improvement is different from breakthrough improvement. Breakthrough requires special methods and leadership support to attain significant changes and results. It also differs from

Table 5.1 Characteristics of systemic failure

Other sectors	Construction industry
Commitments consistently missed	Missed timelines
Late delivery	Project time overrun
Last-minute crunches	
Spiralling costs	Project cost overrun
No management visibility into progress	Lack of proper project management, scheduling
Quality problems	Reworks, quality control
Too much rework	Rework
Functions do not work correctly	Supply chain issues
Customer complaints after delivery	Clients/occupants/end user's dissatisfaction
Poor morale	Poor attitude to work by the labour force

planning and control. Breakthrough requires taking a "step back" to discover what may be preventing the current performance level from meeting its customers' needs. By focusing on attaining breakthrough improvement, leaders can create a system to increase the rate of improvement. By attaining just a few vital breakthroughs year after year (The Pareto Principle), the organisation can outperform its competitors and meet stakeholder needs.

(The Juran Trilogy: Quality Planning | Juran, 1990)

The maturity models are essential to take a process, domain, or industry from a level of immaturity through a systematic and incremental improvement process. Given the present reactionary state of the construction industry based on its failures, it can be said to be immature. Maturity is the potential for growth in capability (Paulk et al., 1993). Maturity models are designed for an entire process or focus areas like the development and maintenance of specific processes or deliverables, alignment with other disciplines, and training of functional domains (Van Steenbergen et al., 2010). It is thus important because maturity models, if adhered to, can determine the quality of system products. The quality of a system is highly influenced by the quality of the process used to acquire, develop, and maintain it (Carnegie Mellon University, 2005).

Given the reactionary nature and late adoption culture regarding technology (WEF & BCG, 2016), a maturity model is required for the technology adoption measurement process. It ensures this process undergoes a systematic and incremental adoption of technology from one maturity level to the next. It will provide a diagnosis of the industry in terms of BIM adoption, help in prescribing the required maturity levels, and be used as a benchmarking tool. This is in line with the three uses as outlined by Becker and Knackstedt (2009); de Bruin et al. (2005); and Pöppelbuß and Röglinger (2011). A maturity model must fulfil descriptive, prescriptive, and comparative purposes.

5.4   Maturity Assessment Framework Development: Classification

Maturity models are developed to provide a framework for excellence through a systematic, stepwise framework. This framework assesses the current state and offers the necessary guidelines for improvement to achieve a higher level of performance. Maturity models are non-restrictive and can be developed for any process or organisational space that requires improvement. It has been applied to innovation, business processes, new product development, project management, supply chain management, and people capability.

Researchers have adopted maturity-based approaches in developing performance improvement frameworks: maturity models and maturity grids. Two prominent maturity frameworks that deployed these approaches are Capability Maturity Model Integration (CMMI) (maturity model) and Crosby's Quality Management Maturity Grid (QMMG) (this adopted a maturity grid approach). Although the approach might be different, both are employed in the assessment and improvement

Table 5.2 Difference between maturity grid and maturity model (Maier et al., 2006, 2012)

S/N	Aspects	Maturity grid	Maturity model
1	Orientation	This applies to companies in an industry; it is company focused	It is process specific
2	Mode of assessment	The structure shows levels of maturity against KPIs of performance in a cell. These cells normally contain descriptions in text form describing the required performance per level.	Assessment is done by: - Likert or binary yes/no questionnaires - Checklist to assess the performance
3	Intent	They are less complex.	They are more of a complex assessment tool and follow an internationally recognised standard format.

of processes. Both approaches also deploy maturity levels and process areas, but the presentation is one of the key differences. However, according to Maier et al. (2012), maturity grids are preferable by companies because it is cost-effective and time-saving. However, it has not gained popularity like maturity models, especially in academia. The differences between them are stated in Table 5.2.

It is worth of mention that maturity assessment in the construction industry has been through maturity models. None has applied the maturity grid approach.

Maturity models are known to be usually developed to be staged or stepwise in their deployment. This is because they provide a continuous framework whereby the achievement of a level is a prerequisite to the next, as the outputs on that level serve as the input for the next. This was supported by the stage theory by Kazanjian and Drazin (1990). The work tested stages of growth of technological ventures based on a four-stage model. It posits that each stage of growth contributes to higher growth rates. It further says that problems occur sequentially and define the next stage that the business must pass through to achieve viability.

Simply put, it states that the growth rate depends on the match between stages and the structure. According to Gottschalk (2009), the stages are sequential, hierarchical, and not reversible and evolve over organisational structures and processes. Most maturity models are developed according to the outline provided by Solli-Saether and Gottschalk (2010). The development of maturity models follows the perspective of the stage theory; hence, maturity models have maturity levels depicting stages. These stages are sequential and are prerequisites for the achievements of the next.

The development of maturity models follows the perspective of the stage theory; hence, maturity models have maturity levels depicting stages. These stages are sequential and are prerequisites for the achievements of the next (Figure 5.1).

Theoretical work

| 1 Suggested stage model | 2 Conceptual stage model | 3 Theoretical stage model | 4 Empirical stage model | 5 Revised stage model |

Ideas from previous research → Dominant problems for stages → Benchmark variables by theories → Values of benchmark variables

Ideas from practitioners and practice → Case studies to different stages → Focus group discussions → Survey research

Empirical work

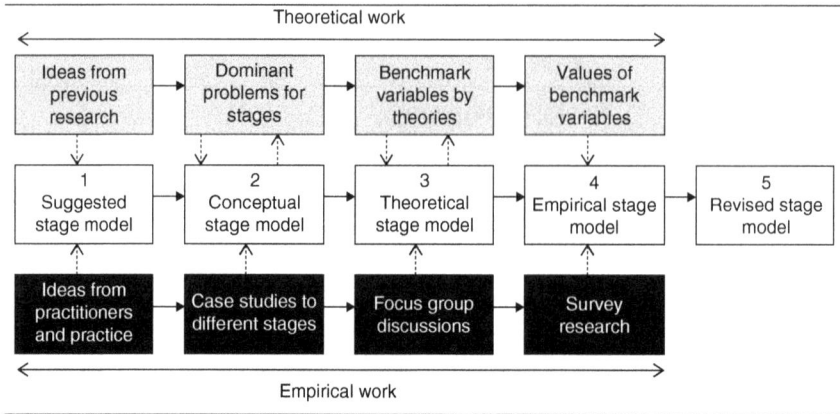

Figure 5.1 Suggested procedure for the stages of the growth modelling process (Solli-Saether & Gottschalk, 2010)

5.5   Steps to Developing a Maturity Model

The development of a maturity assessment model has been variously achieved by different researchers and contexts; however, the approach has been adopted. They are developed to provide a road map and guide in achieving process optimisation. The development of maturity models has been particularly rampant in the digitisation era; for instance, there have been maturity models developed for the industry 4.0, digital transformation, and BIM, among others (Gökalp & Martinez, 2021; Schumacher et al., 2016; Yilmaz et al., 2019).

Generally, maturity models consist of the scope of assessment and the maturity levels. However, to achieve a standardised approach to developing maturity models, studies have proposed steps to achieving a maturity model. According to Pöppelbuß and Röglinger (2011), a maturity model must encompass the following:

- Basic information on the purpose, application domain, differentiation from existing maturity models, the design process, and the empirical validation of the model
- An articulated definition of the model constructs
- Documentation
- Definition of maturity levels and their level of granularity
- Methodology for development

Similarly, Becker and Knackstedt (2009) outlined a list and a model for maturity model development. The procedure model outlines that the model must solve a well-defined relevant problem as its focus and must be articulated to solve a problem. The components outlined by the study are:

i Comparison with existing maturity models
ii Iterative procedure

iii Evaluation
iv Multi-methodological procedure
 v Identification of problem relevance to be solved by the model
vi Problem identification in terms of the application domain, conditions for its application, and the benefits
vii Presentation of results
viii Scientific documentation

Another study on this focus is the one by de Bruin et al. (2005) which posits that the following must be done:

 i Definition of scope
 ii Determination of design for the model
iii Determination of domain components to be measured
iv Test for validity and reliability of developed maturity model
 v Deploy: This is the availability of the model to stakeholders

Adekunle et al. (2022) highlight the following as the components of maturity model developments:

 i Defined scope
 ii Design
iii Development of construct and validation
iv Result presentation
 v Deployment and evaluation
vi Systematic documentation

A critical study of the highlighted studies showed a systematic approach to maturity model development that entails deployment to solve specific problems, an articulated methodology, development of dimensions, validation, and the presentation of results. Also, developed maturity models must be deployed to achieve process optimisation and not redundant. Hence, it is imperative that developed maturity models fill a gap that existing models do not yet fill.

5.6   Summary

Maturity models have become widely accepted for process optimisation after the success of the first maturity model. However, a maturity model must be correctly developed containing the required components. This chapter explored the components and process of developing a maturity model. This is important in order to provide a guide and roadmap to maturity model development. It also provides standardisation of the process. The chapter also provided an overview of the maturity model and the classification. To develop maturity models, the procedures and components must be observed.

References

Adekunle, S. A., Aigbavboa, C., Ejohwomu, O., Ikuabe, M., & Ogunbayo, B. (2022). A Critical Review of Maturity Model Development in the Digitisation Era. *Buildings 2022, Vol. 12, Page 858*, *12*(6), 858. https://doi.org/10.3390/BUILDINGS12060858

Becker, J., & Knackstedt, R. (2009). Developing Maturity Models for IT Management- A Procedure Model and Its Application. *Business & Information Systems Engineering.* https://doi.org/10.1007/s12599-009-0044-5

Carnegie Mellon University. (2005). *Capability Maturity Model ® Integration (CMMI ®) Overview.*

de Bruin, T., Freeze, R., Kulkarni, U., Rosemann, M., Bruin, D., de Bruin, S., Freeze, R., & Carey, W. (2005). Understanding the Main Phases of Developing a Maturity Assessment Model. *16th Australasian Conference on Information Systems*, *109*. http://aisel.aisnet.org/acis2005/109

Gökalp, E., & Martinez, V. (2021). Digital Transformation Maturity Assessment: Development of the Digital Transformation Capability Maturity Model. *International Journal of Production Research.* https://doi.org/10.1080/00207543.2021.1991020

Gottschalk, P. (2009). Maturity Levels for Interoperability in Digital Government. *Government Information Quarterly*, 75–81. https://doi.org/10.1016/j.giq.2008.03.003

Kazanjian, R. K., & Drazin, R. (1990). A Stage-Contingent Model of Design and Growth for Technology Based New Ventures. *Journal of Business Venturing*, *5*(3), 137–150. https://doi.org/10.1016/0883-9026(90)90028-R

Maier, A. M., Eckert, C. M., & John Clarkson, P. (2006). Identifying Requirements for Communication Support: A Maturity Grid-Inspired Approach. *Expert Systems with Applications*, *31*(4), 663–672. https://doi.org/10.1016/j.eswa.2006.01.003

Maier, A. M., Moultrie, J., & Clarkson, P. J. (2012). Assessing Organisational Capabilities: Reviewing and Guiding the Development of Maturity Grids. *IEEE Transactions on Engineering Management*, *59*(1), 138–159. https://doi.org/10.1109/TEM.2010.2077289

Michael Van Sickle, M. (2012). *Transitioning from The Software Capability Maturity Model (SW-CMM®) to the Capability Maturity Model Integrated (CMMI®).* https://doi.org/10.1016/j.jaci.2012.05.050

Paulk, M. C., Curtis, B., Chrissis, M. B., & Weber, C. V. (1993). *Capability Maturity Model SM for Software, Version 1.1.* http://www.rai.com

Paulk, M. C., Weber, C. V., & Chrissis, M. B. (1999). *The Capability Maturity Model: A Summary. The Journal of Defense Software Engineering*, 4, 15–17.

Pöppelbuß, J., & Röglinger, M. (2011). What Makes a Useful Maturity Model?: A Framework of General Design Principles for Maturity Models and Its Demonstration in Business Process Management. *ECIS 2011 Proceeding*, 28. http://aisel.aisnet.org/ecis2011/28

Proença, D., Vieira, R., & Borbinha, J. (2016). A Maturity Model for Information Governance. *Research and Advanced Technology for Digital Libraries, Tpdl 2016*, *9819*(June), 465–467. https://doi.org/10.1007/978-3-319-43997-6

Schumacher, A., Erol, S., & Sihn, W. (2016). A Maturity Model for Assessing Industry 4.0 Readiness and Maturity of Manufacturing Enterprises. *Procedia CIRP*, *52*, 161–166. https://doi.org/10.1016/j.procir.2016.07.040

Solli-Saether, H., & Gottschalk, P. (2010). The Modeling Process for Stage Models. *Journal of Organizational Computing and Electronic Commerce*, *20*(3), 279–293. https://doi.org/10.1080/10919392.2010.494535

Succar, B. (2010). Building Information Modelling Maturity Matrix. In *Handbook of Research on Building Information Modelling and Construction and Construction*

Infromatics: Concepts and Technologies (pp. 65–103). https://doi.org/10.4018/978-1-60566-928-1.ch004

The Juran Trilogy: Quality Planning | Juran. (1990). https://www.juran.com/blog/the-juran-trilogy-quality-planning/

Van Steenbergen, M., Bos, R., Brinkkemper, S., Van De Weerd, I., & Bekkers, W. (2010). The Design of Focus Area Maturity Models. *CEUR Workshop Proceedings, 662,* 17–19.

WEF, & BCG. (2016). *Shaping the Future of Construction a Breakthrough in Mindset and Technology.* World Economic Forum.

Yilmaz, G., Akcamete, A., & Demirors, O. (2019). A Reference Model for BIM Capability Assessments. *Automation in Construction, 101*(January 2018), 245–263. https://doi.org/10.1016/j.autcon.2018.10.022

6 Evaluation of Relevant Existing Maturity Models

6.1 Introduction

To develop the building information modelling (BIM) maturity model for developing countries, existing maturity models, though not exhaustive, were reviewed. Maturity models from the manufacturing sector, IS, and existing BIM maturity models were critically reviewed. Some of these maturity models are discussed to give a better understanding of existing maturity models. The discussed maturity models are the Capability Maturity Model, Industry 4.0 Maturity Model, and SPICE.

6.1.1 Capability Maturity Model (CMM)

The Capability Maturity Model (CMM) presents a systematic roadmap for implementing the vital practices of an organisational process. The CMM is a reference model of mature practices in a specified discipline used to improve and appraise a group's capability to perform that discipline (Carnegie Mellon University, 2005). It was developed in the 80s (Paulk et al., 1991) by Software Engineering Institute (SEI) at Carnegie Mellon University and has since become the basis for developing several maturity models. CMMs give a framework that identifies standardised process improvement hierarchy, which delivers significant business benefits.

The CMM was designed to guide software organisations in selecting process improvement strategies (Paulk et al., 1991). It is considered to be the earliest maturity framework that was developed. Thus, it formed a basis for many maturity frameworks. However, the CMM is not an all-encompassing framework as it was designed for the software management process and did not cover human resource management (Paulk et al., 1993). The CMM provides five levels of maturity; this is in consonance with the Humphrey's (1987) study (from initial, often chaotic, state to repeatable, then defined, managed, and finally optimising). Maturity levels are measured by achieving specific and generic goals that apply to each predefined set of process areas (Gy¨ & Schuster, 2012).

Organisations that employed the CMM have been reported to have improved business productivity and achieved higher progress towards their business objectives (Herbsleb et al., 1994). Thus, the CMM can be said to achieve its

DOI: 10.1201/9781003373919-8

primary objective. The primary aim of designing the CMM is to guide "software organisations in selecting process improvement strategies by determining current process maturity and identifying the few issues most critical to software quality and process improvement" (Paulk et al., 1993). Due to the flexible nature of the CMM, it has been adapted in different sectors like Systems Engr CMM, EIA 731, People CMM, Acq C MM, IPD CMM, ISO 15504, and Systems security Engr CMM; however, all these developed CMMs had different structures, formats, terms, and ways of measuring maturity (Carnegie Mellon University, 2005).

The Capability Maturity Model has five levels of maturity. The maturity level consists of practices related to a predefined set of process areas capable of improving the organisation's performance. Each maturity level achieves new capabilities that the organisation did not have in the previous level. It thus represents a new level of the organisation that is enabled by the transformation of an organisation's process domains (Curtis et al., 2009). This model is employed to identify, measure, and make process changes to the strengths and weaknesses of an organisation. It forms the basis of many BIM maturity models, including the NBIMS.

The CMM was later metamorphosed into Capability Maturity Model Integration (CMMI). The CMMI presents an integrated set of processes for organisations. Some of the misconceptions about the CMMI are that it was referred to by many as a process architecture when it is a model and that organisation processes must align with the CMMI (Michael Van Sickle, 2012). Most of the identified challenges inherent in CMM were addressed by the CMMI. The improvements in CMMI include the introduction of system engineering practice into the model and the breaking out of analysis and measurement into separate process areas. The CMMI consists of well-defined stages that correspond to maturity levels; these contain a set of processes that forms the focus of an organisation set out to improve its maturity. Process areas form the building blocks that establish the process capability of an organisation; they are made up of related practices performed collectively to achieve a set of goals (Michael Van Sickle, 2012).

6.1.2 Structure Process Improvement for Construction Enterprises (SPICE)

This maturity model is discussed because it is one of the pioneers and most cited in the construction industry. This maturity model project was embarked upon to replicate the success of the software CMM in the construction industry. Thus, it borrowed many of its underlying concepts from CMM and contextualised them into the construction industry space. The maturity levels in SPICE are observed to be similar to CMMI.

SPICE framework presents organisations with an evolutionary and continuous process improvement roadmap. It also allows organisations to prioritise to achieve process improvement. Per level, there are specified key processes required to achieve the specific level. To achieve the process improvement outlined by the framework, organisations must implement the key processes as outlined per level. Maturity levels cannot be jumped as they are successive and required to achieve the next.

The SPICE project was undertaken at the Salford University and was funded by the Department of Environment, Transport and Regions (DETR). The methodology applied was (i) a questionnaire to the construction professionals, (ii) a CMM-type assessment of a small architectural practice, and (iii) an experts' panel workshop (Sarshar et al., 1999). It also applies the five maturity levels like the CMM (Sarshar et al., 2004). It is a process improvement framework for construction organisations. It is focused on the process of project delivery by construction organisations. SPICE is one of the few maturity models out there that has undergone practical evaluation on real-life projects.

The SPICE consists of five maturity levels; these levels address the processes related to project delivery: tendering, design, and construction executed by construction organisations. SPICE employed three concepts in its framework: process capability, process enablers, and process maturity.

Process capability is predictive because it predicts the outcome of a process before it is executed. This is different from process performance, which is the result of the execution of the process. Thus, process capability is forward-looking, while process performance is historic. Process capability entails expected results and tells the result of a process before it is embarked upon. The SPICE framework employed four process capabilities within the five maturity levels. Process enablers can be said to be the activities that are prerequisites for implementing a key process. Process enablers possess the critical features a key process area requires to yield successful results (Jeong et al., 2004). The possession of process enablers ensures the performance and predictability of the processes. SPICE identifies five process enablers: commitment, ability, verification, evaluation, and activities (Jeong et al., 2004; Sarshar et al., 2004).

Process maturity represents how an organisation can define, manage, measure, and control a specific process (Jeong et al., 2004). The rating of an organisation against the maturity levels is classified as being mature (levels 2 and 3) or immature (level 1). However, some factors are moderate and determine the maturity of organisations; they include environmental and external constraints (Sarshar et al., 2004). External constraints identified by SPICE include economic recessions, new supply chain relationships, and mergers. The maturity levels are initial (1), project management (2), best practice and knowledge management (3), quantitative management (4), and continuously improving (5).

6.2   Industry 4.0 Maturity Model for Manufacturing Industry (I40)

The Industry 4.0 maturity model is discussed due to its industry 4.0 inclination concept. Although it was not developed for the construction industry, it is important to explain the maturity models in this study because the study falls in the digitalisation framework for the construction industry. I40 was developed as an extension to the existing industry 4.0 maturity models but with more emphasis on the organisational aspect (Schumacher et al., 2016). It has eight dimensions: strategy, leadership, customers, products, operations, culture, people, governance, and technology. It has five maturity levels.

6.3   Existing BIM Maturity Models and Tools

With the growing discussion and adoption of BIM, different BIM tools have been developed to achieve BIM adoption. Kassem et al. (2020) evaluated the existing BIM maturity tools and their benefits worldwide to understand their level of adoption, applicability, strengths, and weaknesses, among others. Although the reviewed models and tools cannot be considered exhaustive, the study reviewed public and unpublished tools. The study identified tools and methods; the tools provide a platform where assessment is conducted, for instance, online survey or excel workbook, whereas methods only provide the methodology required to measure maturity. Some tools that are missing from Table 6.1 are reviewed in this chapter.

6.3.1   UK BIM Wedge

A vital contribution to the BIM implementation space is the BIM wedge developed for the UK industry by Mark Bew and Mervyn Richards in 2008 (BIM+, 2019) under the auspices of the BIM industry working group. It depicts different milestones that enable the easy pegging and assessment of current status by organisations.

It presents four levels (level 0–3) of BIM maturity. Each maturity levels consists of the acceptable milestones/technological diffusion level commensurate to the level being measured. The levels can be grouped into CAD, 2D/3D iBIM. Although this cannot be considered a complete maturity model in itself, it can be referred to as a strategy diagram for BIM adoption. Level 0 organisations still employ paper in their information exchange, and the format is in CAD. At level 1, the information format is in 2D/3D CAD, and the file sharing is through file-based collaboration. It implies that computer usage is employed at the basic level. At this level, PAS 1192: 2007 is already governing the data exchange environment. At level 2, 3D models of building information are exchanged among professionals. There are more file-based collaboration and library management at this stage as compared to the previous stage, level 3. IFC/IFD are engaged to achieve data integration and interoperability. Web-based servers are involved in storage and working platforms. It is worthy to note that the BIM wedge was an important model in implementing BIM in the United Kingdom. The BIM wedge, a foremost BIM standard in the construction industry, laid the foundation for many BIM studies in new adopters of BIM worldwide. However, it cannot be referred to as a BIM maturity model for the construction industry; it served as a guide for BIM studies and learning. This is because it lacks the necessary components of a BIM maturity model; this includes key process areas and capability levels. The BIM wedge is majorly backed up by the PAS 1192-5, a standard guiding BIM implementation. This has been replaced by the ISO 19650 (UKBIM alliance et al., 2019). The new standard is focused on the information management capability of BIM and other related issues. The previous standard, PAS 1192, has the following components: BS 1192: 2007 (collaboration methodology for managing production, distribution, and the quality of building information), BS 1192-2 and BS 1192-3 (for level 2, they explain and

Table 6.1 BIM maturity and benefits: tools and methods

Tool	Owner	Type	Application
BIM Excellence Online Platform	Change Agents AEC	Maturity tool	Organisation, Project
BIM Online Maturity Assessment	National Federation of Builders (NFB)/CITB	Maturity tool	Organisation
BIM Supporters' BIM Compass	BIM Supporters	Maturity tool	Organisation
CPIx BIM Assessment Form	Construction Project Information Committee	Maturity tool	Organisation
Maturity Matrix: Self-Assessment Questionnaire	Project 13 – Institute of Civil Engineers	Maturity tool	Organisation
NBIMS Capability Maturity Model	National Institute of Building Sciences	Maturity tool	Organisation
Organisational BIM Assessment	Pennsylvania State University	Maturity tool	Organisation
SFT's BIM Compass	Scottish Futures Trust	Maturity tool	Organisation
Supply Chain BIM Capability Assessment	Wates	Maturity tool	Organisation
Vico BIM Scorecard	Vico Software (now part of Trimble)	Maturity tool	Organisation
BIM Maturity Assessment Tool (BMAT)	University of Cambridge	Maturity tool	Project
BIM Maturity Measure	ARUP/Institute of Civil Engineers	Maturity tool	Project
BIM Working Group BMAT	Public Sector Working Group	Maturity tool	Project
Dstl BIM Maturity Assessment Tool	Dstl	Maturity tool	Project
VDC Scorecard	Centre for Integrated Facility Engineers, Stanford University	Maturity tool	Project
Owner's BIMCAT (Competency Assessment Tool)	Giel et al. (2015)	Maturity method	Organisation
BIM Maturity Assessment Tool	Department for Transport	Maturity method	Organisation
Organisational BIM Assessment Profile	Pennsylvania State University	Maturity method	Organisation
BIM Return on Investment Tool	Scottish Futures Trust	Benefits tool	Projects
BIM Value	NATSpec	Benefits tool	Organisation, Projects
BIM Benefits	University of Cambridge	Benefits tool	Projects
BIM Level 2 Benefits Management	PricewaterhouseCoopers Strategy	Benefits method	Projects
TfL BIM Benefits Management Strategy	Transport for London	Benefits method	Projects

Source: Adapted from Kassem et al., 2020.

Table 6.2 Focus of reviewed BIM models

Model	Focus
NBIMS MM and VDC Scorecard	Evaluating BIM Maturity of Construction Projects
BIM CAREM, BIM QuickScan, Organisational BIM AP, BIM MM, and Multifunctional BIM MM	BIM maturity of organisations

specify information management requirements), BS 1192-4 (this defines the use of COBie throughout the facility life cycle for information exchange), and BS 1192-5 (this addresses the security issues in the digital built environment) (Yilmaz, 2017).

Generally, existing BIM models studied are organisation and project tailored. They are also suitable for the developed countries, like the United Kingdom (UK) (Dakhil et al., 2015); they are not suitable for BIM infant industries (Himal & Chitra, 2013). Table 6.2 summarises existing BIM maturity models in the construction industry and their focus.

However, most of the reviewed BIM maturity models are organisation tailored. There are other attributes of the reviewed BIM models. Table 6.3 gives an insight into the attributes of the reviewed existing BIM maturity models. This study adopted the outline by Becker and Knackstedt (2009) to achieve an in-depth review of these existing BIM maturity models. According to Becker and Knackstedt (2009), there are eight steps required to develop maturity models. These eight steps are as follows: Comparison with existing maturity models, iterative procedure, evaluation, multi-methodological procedure, identification of problem relevance, problem definition, targeted presentation of results, and scientific documentation. These steps were adopted by this study for the content analysis of the BIM maturity models. Thus, the evaluation shall discuss these existing maturity models by looking for this information in them.

6.3.2 BIM Capability Assessment Reference Model (CAREM)

Yilmaz (2017) developed this maturity model and seemed to be the latest in the published works on BIM maturity models. Its development is deeply rooted in the meta-model of the ISO/IEC 330xx family. It employed a 4-point rating scale. The purpose was to develop a BIM maturity model suitable for assessing the AEC/FM processes of the facility life cycle phases. It was updated iteratively through expert reviews and an explanatory case study. According to Yilmaz et al. (2019), it consists of two parts: the BIM PRM (provides the definition of AEC/FM processes consisting of process purpose, base practice, process outcomes, and work products) and BIM MF. On the contrary, BIM MF enables BIM capability assessment by including a schema comprising BIM capability levels, associated BIM attributes, and a rating scale (Yilmaz, 2017). Although this is the most recent in the reviewed existing models, it is yet to be evaluated and tested in real-life projects.

This model adopted the following capability levels:

1 Level 0 (Incomplete BIM)
   At this level, BIM is considered non-existent as the expected outcome due to its adoption is yet to be felt. The model refers to this level as a level of BIM non-existent/partial implementation.

2 Level 1 (Performed BIM)
   At this stage, the purpose of BIM implementation is to achieve base practice and standalone BIM outcomes. This level lacks BIM collaboration, data exchange, and integration into the facility life cycle. The BIM attributes of this level are performing BIM (which is the extent of achieving the defined BIM outcomes) and BIM skills. The assessment indicators for Attribute 1, which show that the process is performed at the stage, are BIM work products such as 3D models and quantity take-offs. Also, the BIM authoring tools (Autodesk Revit, tesla structures, etc.) are defined at this stage. For Attribute 2, training budgets, employee BIM certifications, and BIM consultancy are evidenced as a sign of BIM skills achievement.

3 Level 2 (Integrated BIM)
   This level heralds collaboration between project stakeholders while leveraging on the successful implementation of the previous stage. At this level, data exchange is done throughout the facility life cycle phases. This level has two BIM attributes: BIM collaboration (this measures the extent of BIM collaboration and information exchange between the facility life cycle) and interoperability (this is a measure of the extent of data flexibility and interoperability between software applications supported).
   The measurement of attributes at this level is as follows:

   – Attribute 1: The indicators for this attribute include generic work products; for instance, BIM execution plans, customised standards, collaboration tools, and the usage of collaboration data environment are defined.
   – Attribute 2: The indicators for this attribute include BIM work products and resources; for instance, interoperable formats (IFC) are defined.

4 Level 3 (Optimised BIM)
   At this level, the previously integrated BIM is consolidated upon at an enterprise level and continuously improved to achieve the organisation's targets. The BIM attributes for this level are corporate-wide BIM deployment, which is the measure of BIM diffusion among team members and its usage in the facility life cycle; and continuous BIM improvement, which is a measure of the extent to which changes to the BIM practices are planned through the causes of BIM usage variation and investigation of innovative BIM approaches.
   Attribute definition for this level goes thus:

   – Attribute 1 involves the definition of BIM work products employed on construction sites (synchronisation model, libraries of custom BIM projects, and

generic resources, e.g., tablets, sensors, and international standards as assessment indicators).
– The basic definitions for this capability level are on strategy, innovation, and generic resources.

The iterative nature of BIM CAREM is well documented and is scientifically verifiable. It is, however, worthy of note that this model was developed in the Turkish construction industry. Also, the major delineation between this model and BIM maturity models before is that it is rooted in the ISO/IEC 330xx standards.

6.3.3 BIM Maturity Model (BIM3)

The building information modelling maturity matrix (BIM3) was developed by Succar (2010). It is one of the most cited BIM models (Himal & Chitra, 2013). It was developed to assess individual/team competency organisational capability.

BIM3 is flexible and can be employed by organisations or projects. According to Wu et al. (2017), it was developed to overcome the deficiencies in NBIS BIM and BIM proficiency matrix. Due to this, this study did not discuss them. These two were criticised for high subjectivity, limited measurement scope in technical aspects, and inadequate reliability and consistency (Succar, 2009). BIM3 has three main process areas and a five-level scale. Although it was adjudged to be flexible, Wu et al. (2017) opined that it lacks a user guide; thus, its applicability is reduced. It is noteworthy that BIM3 has undergone several improvements since the first time it was first developed, and it is continuously tested in organisations and countries.

BIM3 is referred to be comprehensive comparatively to the ones before it but regarded to be weak in the aspects of information management (Dib & Chen, 2012). This highlighted weakness is the main crux of BIM adoption and usage (UKBIM alliance et al., 2019). Succar's work is also based on the CMM. Its five identified stages are Ad-hoc, Defined, Managed, Integrated, and Optimised. The maturity levels reflect the extent of BIM abilities, outcomes, and requirements as opposed to minimum abilities represented by capability stages (Succar, 2010).

The BIM3 was designed to be attainable, cumulative, specific, flexible, informative, measurable, granular, neutral, relevant, and applicable (Succar, 2010). BIM3 is one of the most cited, evaluated, and constantly improved models. To show this, some of the following studies emanated from it: Kassem et al. (2013); Kassem and Succar (2017); Succar (2008, 2009); Succar et al. (2012, 2013); and Succar and Kassem (2015, 2016).

6.3.4 BIM QuickScan

This maturity model was developed by Berlo et al. (2012). It was launched in Holland in 2011 according to Wu et al. (2017), but Berlo et al. (2012) claimed it was created in 2009. It is a BIM benchmarking tool for companies, and it employs both qualitative and quantitative assessments. It was built for and tested in the Netherlands.

Table 6.3 Key attributes of existing BIM maturity models

Element	BIM CAREM (Yilmaz et al., 2019)	BIM3 (Succar, 2010)	BIM QuickScan (Berlo et al., 2012)	Organisational BIM assessment profile (Pennsylvania state University, 2013)	VDC BIM scorecard (Kam, Senaratna, Mckinney, Xiao, Song, 2013)	NBIS BIM (NIBS, 2015b)	BIM proficiency matrix (Indiana University, 2015)	Multifunctional BIM maturity model (Liang et al., 2016)
Objectives	Develop a formal BIM capability assessment framework for the AEC/FM process	To assess individual/team competency organisational capability	Model for the Netherlands	To assess organisation maturity of BIM, for building owners				To assess BIM maturity in projects, companies, and the industry.
Focus	AEC/FM	AEC/FM	AEC/FM	AEC/FM	AEC/FM	AEC/FM	AEC/FM	AEC/FM
Analysis dimensions	Organisations	Organisation	Organisation	Organisation	Organisation	Project	Project	Industry, organisation, projects
Process areas	Process, technology, organisations, human aspect, BIM standard	Process. People, technology	Organisation and management, mentality and culture, information structure and information flow, tools and application	Strategy, use, process, information, infrastructure, personnel	Portfolio and project management, cost planning, schedule control, production control, coordination, design team management	Data richness, life cycle view, roles/disciplines, change management, business protocol, timelines, delivery method, graphical information, spatial capability information accuracy, IFC	Physical accuracy of the model, IPD methodology, Calculation mentality, Location awareness, Content creation, Construction data, As-built modelling, FM data richness	Technology, process, protocol

	4(0-3)	5	6	6	5	6	6	4
Maturity levels								
Inspiring framework	ISO/IEC 330xx family	CMM						
Assessment method	Exploratory case study, expert review	Not clear, but it provides online assessments	Pilot projects, expert review, statistical tests	Not defined	Pilot projects, user interviews, expert reviews, statistical tests	Pilot projects, user interviews, expert reviews	Not clearly stated	Not clearly stated
Maturity levels	Incomplete BIM Performed BIM Integrated BIM Optimised BIM	Initial Defined Managed Integrated Optimised	0 to 5	Non-existent Initial Managed Defined Quantitatively managed Optimising	Conventional Practice Typical Practice Advanced Practice Best Practice Innovative Practice	Not certified Minimum BIM Certified Silver Gold Platinum	Working towards BIM Certified BIM Silver Gold Ideal	Stage 0 Stage 1 Stage 2 Stage 3

Both individuals and companies have employed it, some within companies, and other assessments were done online. Although it was not designed originally to assess individuals (consultants) (Berlo et al., 2012), it employed four areas or chapters and 44 questions. These four main chapters represent both "hard" and "soft" aspects of BIM (Sebastian & Van Berlo, 2010). The chapters are:

i Chapter 1: Organisation and management (corporate management)

The KPIs are vision and strategy, distribution of roles and tasks, organisation structure, quality assurance, financial resources, and partnership on corporate and project levels.

ii Chapter 2: Mentality and culture (organisational culture)

The KPIs measure BIM acceptance among the staff and workers, group and individual motivation, presence and influence of the BIM coordinator, knowledge and skills, knowledge management, and training.

iii Chapter 3: Information structure and information flow (data structure and information flow)

The KPIs are the use of modelling, open ICT standards, object libraries, internal and external information flow, type of data exchange, and type of data in each project phase.

iv Chapter 4: Tools and applications (technology platforms and tools).

These include the hardware and software-related KPIs, which are the use of a model server, type and capacity of the model server, type of software package, advanced BIM tools, model view definitions, and supporting rules.

BIM QuickScan is used to gain insight into the organisation's strengths and weaknesses concerning BIM usage (Yilmaz et al., 2019). To make use of it, it employs a quantitative evaluation approach with a multiple-choice questions approach. It is a free web-based questionnaire with a rating scale to assess the organisation. A special feature of the web-based assessment is time-bound; it must be performed within one day. Like BIM3, it has also been validated via pilot projects, user interviews, and statistical tests. It employs six levels of BIM maturity.

The BIM QuickScan is well documented, and the iterative process is well documented in Sebastian and Van Berlo (2010).

6.3.5 *Organisational BIM Assessment Profile*

This BIM assessment profile was created by Pennsylvania State University computer-integrated construction (CIC) (Pennsylvania State University, 2013); it provides a guide for facility owners. Still, the focus is on organisation BIM assessment. It comprises 20 measures and six maturity levels. However, its BIM uses coverage in the AEC/FM industry is incomplete (Yilmaz et al., 2019).

To achieve effective integration of BIM within an organisation, it proposes three planning procedures. They are as follows:

I Strategic planning consists of steps to aid proper assessment of an organisation's condition and develop a transition plan for BIM implementation through the alignment of BIM goals and objectives with BIM uses and maturity levels. The purpose of strategic planning is to provide a procedure for facility owners to determine the BIM goals and objectives and establish a framework that enables the accomplishment of the goals and objectives. The steps involved are:

  i Assess the current internal and external levels of BIM integration in the organisation.
  ii Align the organisation's BIM goals through the identification of the desired BIM maturity levels.
  iii Advance the identified BIM maturity level by developing a defined advancement strategy for the organisation.

II Implementation planning follows immediately after the strategic planning stage. It is to develop a detailed implementation plan that fits into the organisation's process. It is also required to determine and document an articulated implementation guideline and protocol. This plan includes the following steps:

  i Process maps: This outlines the integration of BIM into the organisation practice. It details the "how" by detailing the procedure required.
  ii Information: This details the requirements that support the BIM implementation in the organisation.
  iii Technology infrastructure supporting the process
  iv Education and training are required for the personnel in charge of BIM operation and other resulting data.

III Procurement planning is done to identify the major issues central to creating BIM contract requirements before the inception of a facility project by the owner. Its purpose is to enable all stakeholders to be on the same understanding regarding the BIM needs and requirements. It ensures that the facility owner's needs are met and the successful implementation of BIM is achieved throughout the project life cycle. The core components are:

  i Team selection criteria
  ii Contract requirements
  iii Standard BIM project execution plan template

IV This assessment model identified six elements necessary for BIM planning. The BIM elements defined according to the Pennsylvania state (2013) are as follows:

  i Strategy: This defines BIM goals and objectives, assesses change readiness, and considers the management and available resource support.

ii Information: This identifies the methods through which BIM can be implemented for generating, processing, communicating, executing, and managing information about the owner's facilities.

iii Process: This gives a full description of how to accomplish the BIM uses by documenting the current methods, designing new processes leveraging BIM, and developing the plans required for a smooth transition.

iv Information: This contains the information needs of the organisation, including the model element breakdown, level of development, and facility data articulately.

v Infrastructure: This determines the technology infrastructure to support BIM, including computer software, hardware, networks, and physical workspaces.

vi Personnel: This establishes the roles, responsibilities, education, and training of the active participants in the BIM processes established.

6.3.6 *Virtual Design Construction BIM Scorecard*

Virtual and design construction (VDC) scorecard was created at Stanford University. It was developed in 2009 and revised in 2011 (Giel et al., 2012), but (Wu et al., 2017) claimed it was developed in 2012. The purpose was to provide a holistic, practical, and adaptive approach to BIM evaluation. The scoring used by the VDC scorecard covers four major areas of VDC, which are Planning, Adoption, Technology, and Performance (Kam, Senaratna, Mckinney, Xiao, Song, 2013). It has 27 questions and employs five capability levels. According to Kam, Senaratna, Mckinney, Xiao, Song et al. (2013), VDC has ten divisions and fifty-six measures, and the confidence level is measured by seven factors to assess maturity while assessing the maturity of BIM implementation on projects. The tool has several distinct features, such as establishing confidence level, which analyses input data and quantitative measurements of the degree of objective compliance (Wu et al., 2017). However, assessing the achievements of performance targets in a progressive manner is difficult due to the large number of quantitative and qualitative measures employed (Kassem et al., 2020).

The development of VDC is based on four criteria. They are as follows:

i Holistic: Unlike previous frameworks, it goes beyond capturing the performances of implementing BIM on projects. The VDC also provides an improvement in project performance.

ii Quantifiable: The measures employed in this framework ensure monitoring and tracking of the project's progress and BIM maturity. It is believed that quantifiable measures provide accurate and actionable recommendations required for decision-making.

iii Practical: According to VDC developers, prior frameworks are too theory-oriented and thus are difficult to adopt by the practitioner. Hence, the measures deployed by VDC are actionable for the professionals and relates these to industry standard.

iv Adaptive: It was built to be flexible and accommodates the dynamic nature of the industry and the complexities that are project-specific.

The VDC scorecard and BIM QuickScan are different mainly because the BIM QuickScan is organisation oriented while the VDC scorecard is project tailored. The VDC also employed a percentage scoring system.

VDC validation was achieved through its application on 108 unique projects consisting of 11 facility types in 13 countries, resulting in over 150 evaluations. Presently, the VDC scorecard is offered online via BIMscore (NIBS, 2015a). BIM-score is an interactive platform for organisational BIM maturity assessment. It evaluates BIM maturity across the identified four areas by VDC.

The four areas of VDC are defined according to Kam, Senaratna, Xiao, Song (2013) as follows:

i Planning: This aligns defined quantitative and qualitative project objectives and goals with the desired business outcomes. It also identifies the various standards, technologies, and resources relevant to the project's execution.
ii Technology: It evaluates the models and analysis employed by assessing the maturity of the models, the level of detail across project phases, and the success of integration across technologies.
iii Adoption: It assesses the procedures and processes involved in VDC by evaluating the success in aligning stakeholders' talents, motivations, incentives, and business structures to create integrated teams and processes that support the project objectives across all project phases.
iv Performance: This is the assessment of project attainment objectives both quantitatively and qualitatively.

6.3.7 Multifunctional BIM Maturity Model

The multifunctional BIM maturity model was developed by Liang et al. (2016). It was developed in Asia, focusing on the Hong Kong and mainland China experts providing the required feedback at the Delphi and interview stage. It comprises 3 domains, 21 sub-domains, and 4 (0-3) maturity levels (stages). It focuses on the domains of technology, process, and protocol. It was developed to evaluate BIM maturity in projects, companies, and the industry. The other details, like validation methods, are not clearly defined; thus, there is limited information about this BIM maturity model.

The multifunctional BIM maturity model was developed through a four-step reiterative process. This model was tailored after the BIM wedge, especially in its shape. This model used domains instead of process areas. It defines the three domains as follows:

i Technology: It is defined as the collection of BIM skills, methods, techniques, and processes, for instance, software and hardware. This has the following sub-domains: information accuracy, model data, quality assurance and quality control, data security and saving, technology infrastructure needs, BIM elements, and spatial coordination.
ii Process: It is the business process required to generate and influence building data required for design, construction, and operation projects. The sub-domains

for this domain are the clash analysis process, data exchange, CAD/BIM work-flow, cross-disciplinary model coordination, delivery method, BIM project objective, and management support.

iii Protocol: It includes the written and unwritten contractual documents that take precedence over existing agreements; this includes model ownership issues, model requirements issues, and model management issues. The sub-domains are interoperability/IFC support, project deliverables, Doc and modelling stand-ards, standard operating process, role and responsibility, compensation expecta-tions, BIM, and facility data requirements.

The three domains deployed in this model share the same perspective as some other models that view BIM beyond being just a technology.

6.4   Summary

According to the components of maturity model development, reviewing exist-ing maturity models is very important. This chapter presents a review of existing maturity models. It reviewed non-BIM maturity models and BIM maturity models. It provided a basis for understanding the development of existing maturity models. It also helps to understand the component of each reviewed model. The reviewed models were developed for different contexts; however, it is worthy of note that most of them employ a progressive approach to attaining maturity. Each of the maturity models has a defined scope or focus and are designed to fill gaps not yet filled by existing maturity models. However, they all respond to the problem to be solved and the context.

References

Becker, J., & Knackstedt, R. (2009). Developing Maturity Models for IT Management-A Procedure Model and Its Application. *Business & Information Systems Engineering*. https://doi.org/10.1007/s12599-009-0044-5

Berlo, L. van, Van Berlo, L., & Hendriks, H. (2012). BIM Quickscan: Benchmark of BIM Performance in the Netherlands. *CIB W78 2012: 29th International Conference*, 17–19. https://www.academia.edu/1905785/BIM_quickscan_benchmark_of_BIM_performance_in_the_netherlands (Accessed: 2 September 2019)

BIM+. (2019). *Explaining the Levels of BIM*. DIigital Construction Resource. http://www.bimplus.co.uk/analysis/explaining-levels-bim/

Carnegie Mellon University. (2005). *Capability Maturity Model ® Integration (CMMI ®) Overview*.

Curtis, B., Hefley, B., & Miller, S. (2009). *People Capability Maturity Model (P-CMM) Version 2.0, Second Edition*. http://www.sei.cmu.edu

Dakhil, A., Alshawi, M., & Underwood, J. (2015). BIM Client Maturity: Literature Review. *12th International Post-Graduate Research Conference 2015 Proceedings*, MediaCityUK, *June 10–12*, 229–238.

Dib, H., & Chen, Y. (2012). A Framework for Measuring Building Information Modeling Matu-rity Based on Perception Of Practitioners and Academics outside the USA. *Proceedings of the CIB W78 2012: 29th International Conference – Beirut, Lebanon, 17–19 October*.

Giel, B., Issa, R. R. A., & Asce, F. (2015). Framework for Evaluating the BIM Competencies of Facility Owners. *Journal of Management in Engineering, 32*(1), 04015024. https://doi. org/10.1061/(ASCE)ME.1943-5479.0000378

Giel, B., Issa, R. R. A., & Liu, R. (2012). Perceptions of Organizational BIM Maturity Variables in The US AECO Industry. *Proceedings of the CIB W78 2012: 29th International Conference, July,* 17–19. http://itc.scix.net/cgi-bin/works/Show?w78-2012-Paper-69

Schuster, G. (2012). *CMM Capability Maturity Model* (pp. 1–12).

Herbsleb, J. D., Carleton, A., Rozum, J., Siegel, J., & Zubrow, D. (1994). *Benefits of CMM-Based Software Process Improvement: Executive Summary of Initial Results.* http:// repository.cmu.edu/sei

Himal, S. J., & Chitra, W. (2013). Assessing the BIM Maturity in A BIM Infant Industry. *The Second World Construction Symposium 2013: Socio-Economic Sustainability in Construction 14–15 June 2013, Colombo, Sri Lanka,* 62–69.

Humphrey, W. S. (1987). Characterizing the Software Process: A Maturity Framework. *IEEE Software, 5*(2). https://doi.org/10.1109/52.2014

Indiana University. (2015). *Building Information Modeling (BIM) Guidelines and Standards for Architects, Engineers, and Contractors.*

Jeong, K., Siriwardena, M., Amaratunga, R., Haigh, R., & Kagioglou, M. (2004). *Structured Process Improvement for Construction Enterprises (SPICE) Level 3: Establishing a Management Infrastructure to Facilitate Process Improvement at an Organisational Level.* http://usir.salford.ac.uk/9965/

Kam, C., Senaratna, D., Mckinney, B., Xiao, Y., & Song, M. (2013). *The VDC Scorecard: Formulation and Validation.* Center for Integrated Facility Engineering: Stanford University.

Kam, C., Senaratna, D., Mckinney, B., Xiao, Y., Song, M., & Mckinney, B. (2013). *The VDC Scorecard: Evaluation of AEC Projects and Industry Trends.* https://stacks.stanford.edu/ file/druid:st437wr3978/WP136.pdf

Kassem, M., Li, J., Kumar, B., Malleson, A., Gibbs, D. J., Kelly, G., & Watson, R. (2020). *Building Information Modelling: Evaluating Tools for Maturity and Benefits Measurement.* Centre for Digital Built Britain, 184.

Kassem, M., & Succar, B. (2017). *Macro BIM adoption: Comparative Market Analysis. 81*(September 2016), 286–299. https://doi.org/10.1016/j.autcon.2017.04.005

Kassem, M., Succar, B., & Dawood, N. (2013). A Proposed Approach to Comparing the BIM Maturity of Countries. *Proceedings of the CIB W78 2013: 30th International Conference, Succar.* http://novaprd-lb.newcastle.edu.au/vital/access/manager/Repository/uon:15200

Liang, C., Lu, W., Rowlinson, S., & Zhang, X. (2016). Development of a Multifunctional BIM Maturity Model. *Journal of Construction Engineering and Management, 142*(11). https://doi.org/10.1061/(ASCE)CO.1943-7862.0001186

Michael Van Sickle, M. (2012). *Transitioning from The Software Capability Maturity Model (SW-CMM®) to the Capability Maturity Model Integrated (CMMI®).* https://doi. org/10.1016/j.jaci.2012.05.050

NIBS. (2015a). National BIM Standard-United States. *BuildingSMARTalliance, December,* 1–15. www.nationalbimstandard.org

NIBS. (2015b). National BIM Standard - United States ® Version 3_5.2 Minimum BIM. *National Institute of Building Sciences, BuildingSMART Alliance,* 1–13. https://www. nationalbimstandard.org/files/NBIMS-US_V3_5.2_Minimum_BIM.pdf

Paulk, M. C., Chrissis, M. B., & Weber, C. V. (1991). *The Capability Maturity Model for Software.* Carnegie-Mellon Univ, Pittsburgh, PA, Software Engineering Inst.

Paulk, M. C., Curtis, B., Chrissis, M. B., & Weber, C. V. (1993). *Capability Maturity Model SM for Software, Version 1.1.* http://www.rai.com

Pennsylvania state University. (2013). BIM Planning Guide for Facility Owners. In *The Pennsylvania State University, University Park, PA, USA: Vol. Version 2.* http://bim.psu.edu

Sarshar, M., Finnemore, M., Haigh, R., & Goulding, J. (1999). SPICE: Is a Capability Maturity Model Applicable in the Construction Industry? In *International Conference on Durability of Building Materials and Components* (Vol. 30), 2836–2843.

Sarshar, M., Haigh, R., & Amaratunga, D. (2004). Improving Project Processes: Best Practice Case Study. *Construction Innovation, 4*(2), 69–82.

Schumacher, A., Erol, S., & Sihn, W. (2016). A Maturity Model for Assessing Industry 4.0 Readiness and Maturity of Manufacturing Enterprises. *Procedia CIRP, 52*, 161–166. https://doi.org/10.1016/j.procir.2016.07.040

Sebastian, R., & Van Berlo, L. (2010). Tool for Benchmarking BIM Performance of Design, Engineering and Construction Firms in the Netherlands. *Architectural Engineering and Design Management, 6*(SPECIAL ISSUE), 254–263. https://doi.org/10.3763/aedm.2010.IDDS3

Succar, B. (2008). Building Information Modelling Framework: A Research and Delivery Foundation for Industry Stakeholders. *Automation in Construction, 18*, 357–375. https://doi.org/10.1016/j.autcon.2008.10.003

Succar, B. (2009). Building Information Modelling Framework: A Research and Delivery Foundation for Industry Stakeholders. *Automation in Construction, 18*(3), 357–375. https://doi.org/10.1016/j.autcon.2008.10.003

Succar, B. (2010). Building Information Modelling Maturity Matrix. In *Handbook of Research on Building Information Modelling and Construction and Construction Infromatics: Concepts and Technologies* (pp. 65–103). https://doi.org/10.4018/978-1-60566-928-1.ch004

Succar, Bilal & Agar, Carl & Beazley, Scott & Berkemeier, Paul & Choy, Richard & Giangregorio, Rosetta & Donaghey, Steven & Linning, Chris & Macdonald, Jennifer & Perey, Rodd & Plume, Jim. (2012). BIM Education, BIM in Practice. Australian Institute of Architects.

Succar, B., & Kassem, M. (2015). Macro-BIM adoption: Conceptual Structures. *Automation in Construction, 57*, 64–79. https://doi.org/10.1016/j.autcon.2015.04.018

Succar, B., & Kassem, M. (2016). Building Information Modelling: Point of Adoption. *CIB World Congress, Tampere Finland, May 30 – June 3, 2016, 1*(2016), 1–11.

Succar, B., Sher, W., & Williams, A. (2013). An Integrated Approach to BIM Competency Assessment, Acquisition and Application. *Automation in Construction, 35*, 174–189. https://doi.org/10.1016/j.autcon.2013.05.016

UKBIM alliance, CDBB, & BSI. (2019). *Information Management According to BS EN ISO 19650.*

Wu, C., Xu, B., Mao, C., & Li, X. (2017). Overview of BIM Maturity Measurement Tools. *Journal of Information Technology in Construction (ITcon), 22*, 35. http://www.itcon.org/2017/3

Yilmaz, G. (2017). *BIM-CAREM: A Reference Model for Building Information Modelling Capability Assessment.* The middle east technical University.

Yilmaz, G., Akcamete, A., & Demirors, O. (2019). A Reference Model for BIM Capability Assessments. *Automation in Construction, 101*(January 2018), 245–263. https://doi.org/10.1016/j.autcon.2018.10.022

7 Conceptualisation of the BIM Maturity Model

7.1 Introduction

This chapter focuses on the conceptualised building information modelling (BIM) maturity model; it explores the dimensions of the conceptualised model, the maturity levels, and the sub-variables under each dimension. This chapter concentrates on six dimensions: people, process, technology, stakeholder orientation, effective BIM training, and ecosystem regulation. The first three constructs are from the review of existing maturity models, while the latter three are the gaps identified from the review of the existing maturity model. This study posits that the gaps are contextually derived for developing countries and are required to achieve BIM maturity in developing countries. In addition, this chapter identifies the outcomes due to the achieving of an optimum BIM maturity level in developing countries.

7.2 Variable Selection for BIM maturity

Modal constructs for this study imply those domains, measures, dimensions, or process areas used commonly by different studies previously. Heretofore, different researchers chose a nomenclature they deemed fit for their study, although they meant the same thing. Simply put, they are the building blocks to the achievement of capability. According to Wu et al. (2017), the most common measures adopted by previous studies can be classified into five: process, technology, organisation, human, and standard. This study reviewed previous BIM maturity models and the frequently deployed process areas that were established (Table 7.1). It was found that the commonly used measures are process, technology, organisation, info/ data governance, and human/ resource management. Some of the constructs are explained below.

7.2.1 Process

The BIM process is invented to ensure a quality improvement of the construction process. According to the CMM, a process is defined as "a set of activities, methods, practices, and transformations that people use to develop and maintain

DOI: 10.1201/9781003373919-9

Table 7.1 Development of the modal construct

Process areas		Existing BIM Maturity models								Process areas
		BIM CAREM	BIM3	BIM QuickScan	Organisational BIM assessment profile	VICO BIM scorecard	NBIS BIM	BIM proficiency matrix	Multifunctional BIM maturity model	
1	Process	x	x							Process
4	Technology	x	x							
5	Tools and application			x						Technology
6	Infrastructure				x					
7	Use									
8	Organisations	x								
9	Portfolio and project management					x				Organisation
10	Production control					x				
11	Organisation and management			x						
13	Design team management					x				
14	Information structure and information flow			x						Technology
16	BIM standard	x								
17	Information				x					
18	Construction data								x	
19	As-built modelling								x	
20	FM data richness								x	
21	Human aspect	x								Human and resource management
22	Cost control					x				
23	Cost planning					x				
26	Coordination					x				
27	Schedule planning					x				
29	Personnel				x					
31	People		x							
32	Mentality and culture			x						
33	Physical accuracy of the model									
34	IPD methodology									
35	Calculation mentality									
36	Location awareness									
37	Content creation									

software and the associated products (e.g., project plans, design documents, code, test cases, and user manuals)." A process according to the organisational maturity model (Pennsylvania State University 2013) means to achieve BIM uses through the documentation of the current methods and designing future methods by leveraging on BIM. The process is important because the quality of a system or product is highly influenced by the quality of the process used to acquire, develop, and maintain it (Carnegie Mellon University, 2005). The construction process is, therefore, a set of activities, methods, practices, standards, and innovations employed by stakeholders in developing and maintaining construction products and ancillary services throughout their life cycle. The development of BIM maturity models is geared towards improving the construction process to achieve a productive and more efficient construction process and, in extension, quality products. Succar (2010a), however, observed that most organisations ignore the process aspect and the policy implications of BIM but rather treat it majorly as a technology.

The process dimension is measured in four of the reviewed existing BIM maturity models: BIM CAREM, BIM3, VDC BIM scorecard, and BIM QuickScan (Table 7.2).

Table 7.2 Definition of process in existing BIM maturity models

S/N	Maturity model	Variables for measurement
1	BIM CAREM	Change orders management process through BIM, coordination and handover processes between project phases, interaction coordination and communication among multiple disciplines or stakeholders, information collection and response information flow management, information generation and documentation (e.g., quantity take-offs and weekly schedules), delivery processes of BIM-relating products and services, reuse procedures of BIM-related information and data
2	BIM3	Infrastructure: physical and knowledge-related
		Products and services specification, differentiation, project delivery approach, and R&D
		Human resources: competencies, roles, experience, and dynamics
		Leadership: innovation and renewal, strategic, organisational, communicative, and managerial attributes
3	Organisational BIM assessment profile	Interaction coordination and communication among multiple disciplines or stakeholders
4	VDC BIM scorecard	Project benefit, technology phases, IPD, and integrated meetings
5	Multifunctional BIM maturity model	Clash analysis process
		Data exchange
		CAD/BIM workflow
		Cross-disciplinary model coordination
		Delivery method
		BIM project objective
		Management support

7.2.2 Technology

This construct was employed by six of the reviewed maturity models. The models are BIM3, BIM QuickScan, VDC BIM scorecard, multifunctional BIM model, and organisational BIM assessment profile. This is measured from different perspectives and levels of granularity. The organisational BIM assessment profile (Pennsylvania State University, 2013) was measured as infrastructure that supports BIM, including hardware, software components, and physical workspaces. The study by Yilmaz et al. (2019) measured technology in terms of infrastructure like that for organisational BIM assessment profile; however, it considered other technical aspects of the construct. BIM CAREM also measured this construct through data richness and accuracy, information security, and spatial accuracy. BIM3 measurement variables are similar to that of the organisational BIM assessment profile. It measured technology using software, hardware, and networks (Succar, 2010b; Succar, Sher, et al., 2012). The measurement variables for BIM QuickScan are a combination of infrastructure and information structure and flow (Berlo et al., 2012; Sebastian and Van Berlo, 2010). This is similar to the variables employed in BIM CAREM. It can be observed that the reviewed existing BIM maturity models all employed infrastructure and data/information integrity as the variables to measure technology. Thus, it is concluded that the data and information integrity variables are all classified under the technological construct. According to BIM CAREM, this can be said to be the technical aspect of the construct. Technology can be said to be the infrastructure and all information structure and integrity required for BIM implementation (Table 7.3).

7.2.3 Organisation

BIM CAREM, BIM QuickScan, and VDC BIM scorecard employed this process area in their models. VDS BIM scorecard measured organisation using

Table 7.3 Definition of technology in existing BIM maturity models

S/N	Maturity model	Variables for measurement
1	BIM3	Data richness and accuracy, information security, and spatial accuracy.
2	Organisational BIM assessment profile	Infrastructure (computer software, hardware, networks, and physical workspaces)
3	VDC BIM scorecard	Maturity depth and breadth, coverage (level of detail, model use life cycle), integration (communication, interoperability, HW/SW adequacy)
4	BIM QuickScan	Use of model server, type and capacity of model server, type of software package, advanced BIM tools, model view definitions, and supporting rules.
5	Multifunctional BIM maturity model	Information accuracy, Model data, quality assurance, and quality control, data security and saving, technology infrastructure needs, BIM elements, and spatial coordination

Table 7.4 Definition of organisation in existing BIM maturity models

S/N	Maturity model	Variables for measurement
1	BIM CAREM	BIM missions and objectives at operation level, senior management supports (e.g., personnel and finance), attitude of management and leadership towards BIM, Research, and Development efforts (R&D), objectives establishments, and degree of compliances
2	BIM QuickScan	Vision and strategy, distribution of roles and tasks, organisation structure, quality assurance, financial resources, and partnership on corporate and project level
3	VDC BIM scorecard	Stakeholder involvement, stakeholders attitude, stakeholders action, and number of stakeholders

stakeholder involvement, stakeholder attitude, stakeholder's action, and number of stakeholders. BIM QuickScan referred to process areas as chapters. This process area is measured as organisation and management. According to Sebastian and Van Berlo (2010), the chapters were measured in the form of multiple-choice questions. However, for BIM CAREM, organisational BIM visions, goals, strategies, operational BIM missions and objectives, BIM support and attitude of management and leadership, compliance to establishments' objectives, and research and development were used to measure this construct. It is evident that this construct has a diverse approach by these maturity models; however, BIM CAREM tailored its variables to BIM. This is missing in other maturity models. The organisation is thus the alignment of organisation mission, vision, and objectives with the BIM implementation (Table 7.4).

7.2.4 Human and Resource Management

The human resources domain of the BIM maturity models represents the roles, responsibilities, competency, and training of all active participants involved in the BIM process. This was measured as personnel in the organisational BIM assessment profile (Pennsylvania State University, 2013). Meanwhile, human resources were measured under the process construct in BIM3. The human resource construct in BIM QuickScan was measured in terms of mentality and culture. Unless the workforce practices are supported by organisational behaviour, the improved workforce will not survive (Curtis et al., 2009; Table 7.5).

7.3 Aligning Existing Constructs

From the review of existing maturity models and the prevailing constructs, it is evident that the existing models measured constructs from different perspectives. Some of the models measured the same construct using different variables. Wu et al. (2017) observed that constructs employed in BIM maturity models are generally categorised into five categories: process, technology, organisation, standard,

Table 7.5 Definition of human resource management construct in existing BIM maturity models

S/N	Maturity model	Variables for measurement
1	BIM CAREM	BIM-related staff experiences, skills, and knowledge of BIM staff/stakeholders, Arrangement of BIM-related duties and roles, BIM-related training and education, Existence and functions of BIM champion/leader, Awareness, attitudes, enjoyments, and involvements of employees/stakeholders towards BIM
2	BIM QuickScan	vision and strategy, distribution of roles and tasks, organisation structure, quality assurance, financial resources, and partnership on corporate and project level
3	VDCBIM scorecard	stakeholder involvement, Stakeholders attitude, Stakeholders action, and number of stakeholders
4	Organisational BIM assessment profile	roles, responsibilities, education, and training of the active participants in the BIM process establish

Table 7.6 Aligned constructs

New constructs	Existing
Process/Operations	Processes and activities, Operations, portfolio, and project management
Human resource	The human aspect, organisation, leadership, personnel, participants, people, mentality, and culture
Technology	Data governance, BIM standard, Tools and application, Infrastructure, Use, information structure and flow, construction data, information, as-built modelling, data richness, physical accuracy of model.

and human. In a similar categorisation effort, Yilmaz (2017) observed the common categories: process, stakeholder/personnel, standard, software, hardware, and data.

After reviewing existing BIM models, this study observed that the categorisation deployed by these models seems to be similar. Also, the similarity observed is a pointer to how they motivated one another (Yilmaz, 2017). However, in a bid to be different beyond the deployed maturity levels and nomenclature, they employed different concepts to measure the process areas. Mostly, they split the constructs and measured them under different nomenclature while measuring the same thing. This study reclassified the existing constructs to achieve generally encompassing constructs to have a unified process areas definition. Table 7.6 gives the grouping.

Thus, this study categorised the existing key processes observed in the reviewed BIM maturity models under three domains, as shown in Table 7.6. This study will thus build on the BIM3 by Succar because it is the most evaluated in the industry among the reviewed existing models. For this study, these constructs are rooted in the ISO 19650 series that defines the boundaries for information management using BIM. It was adopted for this study because BS 19650 provides a framework for information management for all actors. Information management includes information exchange, recording, versioning, and organising during an asset life cycle (strategic planning, initial design, engineering, development, documentation and construction, day-to-day operation, maintenance, refurbishment, repair, and end-of-life) irrespective of scale and complexity. According to the standard, actors include the person, organisation, or organisational unit involved in the construction process of a built asset.

– **Processes** for this study are every activity carried out by teams during the life cycle of any building asset to achieve effective information management according to ISO 19650-2:2018. The ISO 19650 series was released to replace the ISO 1192:2007 and thus supersedes it. ISO 19650 provides the principles and concepts for information management using BIM. They are pivotal to the successful execution of the project.

BIM has been established to be more than just technology but also a process change (Eastman, 2011). The study of the process domain is important because of the associated changes. Technology-supported processes take place in a collaborative environment and make use of existing knowledge and experience. The collaborative space involves communication and sharing data without ambiguity, loss, or contradiction. According to BS 1192:2007(British Standards Institute, 2015), these processes include automation of 3D model, data, drawing, and document production processes; quality checking and document comparisons; filtering and sorting; indexing; and searching project materials. These processes apply to parties involved in the management of construction information production and distribution.

The process is rooted in the common data environment (CDE). According to the BS 1192:2007, the CDE involves four phases (Figures 7.1 and 7.2). The CDE is meant to be instituted by the appointing party before the commencement of the project (ISO 19650-2:2018, 2018) BS 19650.

The standard recognises the following roles involved in the information management process:

A   Architect
B   Building Surveyor
C   Civil Engineer
D   Drainage, Highways Engineer
E   Electrical Engineer
F   Facilities Manager
G   Geographical and Land Surveyor
H   Heating and Ventilation Designer
I   Interior Designer
K   Client
L   Landscape Architect
M   Mechanical Engineer

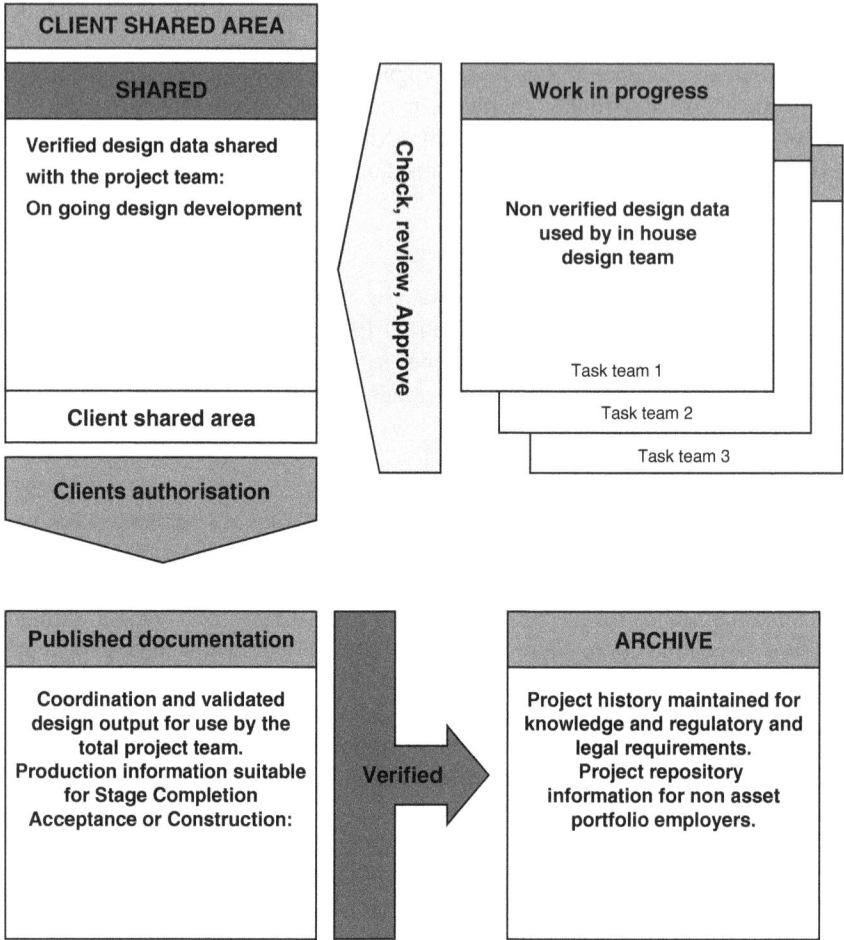

Figure 7.1 Document and data management repository, ISO 1192 (adapted from British Standards Institute, 2015)

P   Public Health Engineer
Q   Quantity Surveyor
S   Structural Engineer
T   Town and Country Planner
W   Contractor
X   Subcontractor
Y   Specialist Designer
Z   General (non-disciplinary)

The BIM processes necessary for effective information management among the roles in a collaborative environment is the focus of the process domain. Reduction

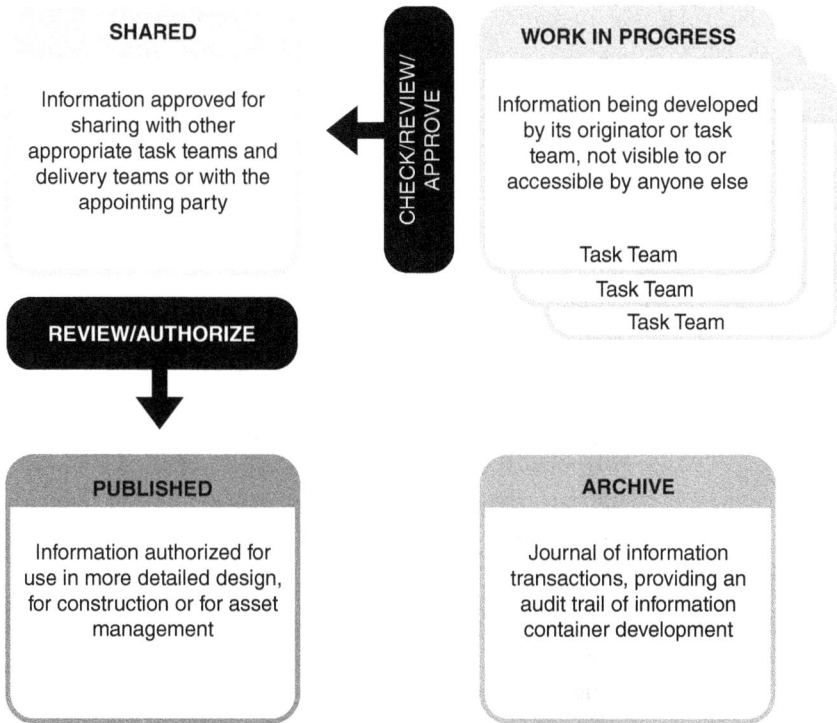

Figure 7.2 Common data environment (CDE) (ISO 19650-1:2018)

of duplicated works, improved information quality, and overall project success are benefits of a collaborative working environment by stakeholders.

- **Technology** includes the hardware, software requirements, and network integration for BIM maturity. This measures the infrastructure required for BIM maturity in an organisation. The prevailing level and its adequacy or otherwise are assessed.
- **People** domain focuses on the roles, perception, organisation culture, responsibilities, skills, and training of BIM actors in an organisation. These actors' areas are outlined in the roles section of BS 1192:2007.

7.4   Gaps from Existing Models

7.4.1 Ecosystem Regulation

The concept of regulation has been studied from a different perspective. Most of the definitions of regulation described it as a restriction of liberty and thus saw it from a negative perspective (Moosa, 2015). However, regulations are enacted to protect stakeholders and promote order in any sector. Regulations take different forms, including setting minimum standards, investigating, and prosecuting misconduct, among others. This is a very important distinction between the human and the animal kingdom; regulations are enacted to encourage order. Different forms of regulation exist, which include:

– Government imposed legal restrictions
– Public standards of expectations
– Licensing
– Rules and conduct of professional bodies

Achieving regulation can either be by the government or pluralised, also known as smart regulation. Smart regulation refers to a flexible perspective to achieving regulation that harnesses government, business, and third parties participation (Drahos, 2017). Smart regulation ensures a holistic approach because the instruments adopted are complementary. It thus looks beyond the regulator (government) and the regulated (business entities) concept. Smart regulation looks at regulation from a perspective that regulation is not the job of the government alone. It discourages the monopolisation of regulation by the government. It promotes the idea that regulation is pluralistic in nature in terms of stakeholder involvement and instrument. Regulations can be applied in two ways: to effect structural order in a sector and to regulate the conduct (behaviour) of stakeholders.

The ecosystem has its origin in the biological sciences. It is a community of interacting organisms and their environment. Mutualism exists likewise in the business ecosystem like the natural habitats, where mutually beneficial relationships exist (Moore & Business, 1996). Generally, it can be said to be a complex interconnected system. Peter and Grivas (2016) describe it as interconnecting and interacting stakeholders who form a system. The BIM ecosystem presents the space where all components coevolve and compliments each other. The BIM ecosystem must have an equal and compatible growth among its components, identified products, people, and processes (Gu et al., 2014). This functional and fully matured BIM ecosystem is only possible when equal, compatible, and complementary growth exists among aspects. The South African construction industry (SACI) can be said to be at the skewed adoption stage. This stage is characterised by an unplanned adoption of a radical innovation without adequate assessment of the complementary aspects (Gu et al., 2014). There are different misalignments among the components; thus, there is severe misalignment. Thus, BIM adoption is far from being achieved due to the disconnection and the lopsided development nature of the components. The components are not developed commensurately; thus, it leads to the inability to enjoy full adoption successfully.

Although the classification of exiting components is people, process, and technology against Gu et al.'s (2014) classification of a product, process, and people (skills), the classifications offered by this study are broad and give comprehensive and extensive coverage.

For this study, an ecosystem is defined as a complex network of interdependent, interconnecting, and interacting components (people, process, and technology) in the construction industry whose activities and relationships determine the full implementation of BIM. These relationships and activities are sustainable. The construction industry ecosystem consists of different players and is faced with different challenges. According to Peter and Grivas (2016), the major challenge facing ecosystems is transformation with the digitisation wave successfully. These transformation challenges are regulations, standards, digitisation, and innovation. However, this study will be focusing on the regulations and standards regulating the ecosystem. This pertains to the contractual components of the BIM ecosystem that spell out the contractual responsibilities of each stakeholder collectively.

Contrary to the position of Gu et al. (2014) that an alignment of process, people, and products is all that is required for a balanced ecosystem, this study opines that alignment can only be possible when the ecosystem has technologically aligned standards and governance. New governance structures must be put in place to allow for the adoption and diffusion of BIM by the stakeholders in the ecosystem. This affects the existing process and protocols majorly to achieve BIM compatibility.

Adoption of new technology can be a failed project irrespective of the perception of the human actors or organisation. This is because they meet stiff resistance from the existing societal-technical regime (including the rules and artefacts that form the existing system's structure) (Geels, 2004; Nelson & Winter, 1983). Prevailing socio-technical regimes most time are deeply rooted as they support the existing system. Hence, most actors are always reluctant to let go of it. They are reluctant to let go majorly because of their investments (Nelson & Winter, 1983). Also, these prevailing regimes do not support the recent innovations as the path-breaking innovations present a deviation from the status quo. Thus, for new innovations to succeed in a particular system, they must be accommodated by the socio-technical regimes. The focus is on the challenges of existing socio-technical regimes to accommodate new innovations. According to Walrave et al. (2018), there must be deliberate efforts for strategies to be developed which allow a seamless adoption by societal stakeholders of the values inherent in path-breaking innovations. This is because path-breaking innovations do not get ready support from the existing structure in the environment (Smith & Raven, 2012); the structure includes policies and infrastructure among others. For instance, a report by Mcgraw Hill observed that structural barriers in the ecosystem are the major barrier to BIM adoption in China (McGraw, 2014). The study reported that the introduction of BIM conflicts with the benefits of traditional stakeholder roles and values. Thus, an integration of the path-breaking innovation and existing structures is required.

7.4.2 Regulation Theory

There are different perspectives to regulation and thus different theories and classifications. These theories explain the theoretical underpinnings of regulation and its dynamics. There are different regulatory theories including the Public Interest

Theory of Regulation, the Capture Theory of Regulation, Special Interest Groups Theory of Regulation, and Enforcement Theory of Regulation. It is, however, worthy of note that these theories are interconnected and overlapping. Two theories are adopted for this gap and are discussed below.

7.4.2.1 Public Interest Theory of Regulation

This theory posits that regulations obey the demand and supply law. Consequently, regulations are seen as a response to public demand for the correction of inefficient practices. It is thus generally believed that regulation is enacted to benefit the public and does not serve vested interest (Moosa, 2015). This theory assumes that the government is the sole enforcer of regulations and thus intervenes accordingly in the public's interest. It can be described as responsive regulation (Drahos, 2017). This theory believes in the collaboration of stakeholders to drive the regulation and its enforcement thereof. It can be implied that a dormant public interest body might lead to non-responsiveness of the government. This theory is premised on the assumption that the regulator is sufficiently informed and possesses the enforcement power (Den Hertog, 1975).

In the construction industry, the government has enacted various regulatory frameworks that have brought about changes in the construction industry. This has been particularly observed in the BIM maturity initiative in some BIM-leading countries.

7.4.2.2 Special Interest Groups Theory of Regulation

A special interest group is an association of formally organised individuals who have shared concerns attempting to influence public policy in their favour. They make conscious efforts to affect government policy for their benefit or causes. Special interest groups are pressure groups, advocacy groups, lobby groups, campaign groups, and interest groups, and, in the construction industry, professional associations and other stakeholder groups. This theory posits that powerful interest groups strive "for the use of the coercive power of the government to introduce rules and regulations that would help their businesses." This theory recognises the ability of organised interest groups to effect desired change and the necessity for political alignment. Unlike the previous theory, regulations are driven by other organised stakeholders aside from the government.

7.4.3 Social Network Analysis and Structuration Theory

Social network analysis describes the nature of social relationships in terms of nodes and ties. Ties represent the connection and links, while the nodes represent the human actors in the network. The links connect the nodes and show the relationship that exists between them. Social networks are made up of structural relationships between individuals or organisations (nodes) connected by diverse links. It is rooted in the network theory, which studies graphs to represent relations between objects.

The type of relationship between actors determines the flow of resources (McGee & Warms, 2013), including information. This is termed a mimetic process of contagion where intangible resources are transferred or exchanged. The social network theory explains the structural relationship in an ecosystem and how the relationship between actors affects the coevolution. These relationships were identified using the ecosystem approach as cooperation, complementarity, interdependence, and interconnections (Scaringella & Radziwon, 2018).

Structures govern the relationship between actors. Structures provide regularity and stability to the actions and behaviours of actors in an ecosystem (Rogers, 1962). The structuration theory by Gidden has the focal lens of viewing structure in two dimensions (Giddens, 1984), which is termed the duality of structure (Rose & Scheepers, 2001). A structure, according to structuration theory, refers to the: "rules and resources recursively implicated in social reproduction; institutionalised features of social systems have structural properties in the sense that relationships are stabilised across time and space" (Giddens, 1984). Quoting Gidden verbatim, he defined structure, system, and structuration as:

1	Structure(s)	"Rules and resources, or sets of transformation relations, organized as properties of social systems."
2	System(s)	"Reproduced relations between actors or collectivities, organized as regular social practice."
3	Structuration	"Conditions governing the continuity or transmutation of structures, and therefore the reproduction of social systems"

The duality of structure explains the dependence on two phenomena – structure and agency. Due to this dualism in their relationship, the behaviour of the actors is constrained and enabled by the existing structure. However, actors are capable of transforming structures (Jones & Karsten, 2003). Through their activities, actors determine structures, as structures do not exist independent of human action. Structure is the activity of human actors reproduced over time and space (Giddens, 1984). Hence, an existing structure can be challenged by human actors to produce a new one over time.

Aligning the position of the existing socio-technical regime and the links of the relationship between actors in the social network analysis and the position of the structuration theory where structures do not exist independent of the actors, it suffices to say that for the existing socio-technical regime to align with a path-breaking innovation, the existing structure constraining the action of the actors and determining the relationships must be altered. The existing rules and regulations (structure) making up the socio-technical regime (system) must be rewritten. According to Eastman et al. (2008), for BIM to be widely adopted, there is a need for the existing model of contracts to be changed.

For this study, the existing standards and laws in existence in the construction ecosystem that resist the collaborative nature of the BIM adoption must be rewritten to support it. These regulations must be rewritten to provide a proper structural relationship among the actors because the value inherent in BIM can only be achieved in a collaborative environment.

7.4.3.1 Achieving Ecosystem Regulation

Regulating the construction industry ecosystem through guiding regulations has always been an uphill task. Ecosystem regulation has been difficult to achieve; the Latham report (Latham, 1994) discussed extensively various bye-laws, regulations, and laws in the construction industry. The report posited that laws are not burdensome or unreasonable but are meant to reinforce good practice. Similarly, in achieving BIM maturity, the regulations governing the construction projects and stakeholder relationships, among others, must be well-tailored towards BIM implementation. Studies have shown that to achieve the BIM maturity model, ecosystem regulation is an important factor. Government regulatory efforts have been shown to have a positive relationship with BIM maturity (Kassem et al., 2015; Louay & Kassem, 2018; Succar, 2010b). The United Kingdom construction industry provides a good example of the positive relationship between BIM maturity and government policies. Other parts of the globe where the public sectors show efforts in promoting BIM adoption are the United States, Europe, Asia, and Australasia (Cheng & Lu, 2015).

On the contrary, the regulatory frameworks and guidelines from interest groups are present in the literature (Succar, 2009; Succar, Sher, & Williams, 2012). Thus, BIM ecosystem regulation is pluralistic. BIM-tailored regulations and tools are essential to encourage stakeholder adoption of BIM and can be either regulator or interest group driven.

7.4.4 Stakeholder Orientation

Stakeholders are actors with great interest in a concept. According to the social network analysis, they are linked and in a relationship. This relationship determines the flow of resources and capital owned by these stakeholders. However, these links or networks are in a constant state of movement (actor-network theory). Stakeholders are important to the diffusion of innovation (innovation diffusion theory). According to Rogers' diffusion theory (Rogers, 1962), categories of stakeholders' adoption of innovation are four. Moving from one stage to the other depends on the orientation and perception of the stakeholders regarding the innovation. According to the diffusion of innovation theory, the orientation of actors is central to the rate of diffusion of an innovation. The theory posits: "an individual or other unit of adoption that has knowledge of, or experience with using, the innovation; another individual or other unit that does not yet know about the innovation, and a communication channel connecting the two units" (Rogers, 1962).

It can thus be inferred that actors are central to the spread of innovation as they act as a source of diffusion. However, an actor can only communicate as much as he knows. Hence, proper orientation of the stakeholder is important to the spread of an innovation. On the contrary, it can be said that the lack of BIM diffusion is due to either inadequate orientation or the absence of BIM knowledge among the stakeholders (Figure 7.3).

According to Rogers, stakeholders' orientation of the innovation is also important because the innovation-decision process starts with the knowledge stage. This stage

COMMUNICATION CHANNELS

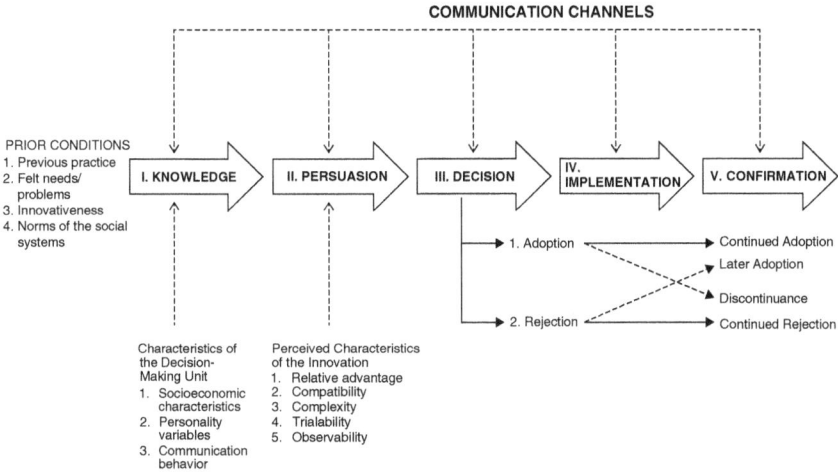

Figure 7.3 Model of stages in the innovation-decision process (Rogers 1962)

is the stage where the actor is exposed to the innovation and how it works. This stage opens the prospective adopter to realise the inherent benefits of the innovation and assist in forming the perception. It also signals the beginning of his adoption journey.

Adopting new technologies has always been a challenge in every sector. To overcome this challenge, many models have been developed: technology acceptance model (TAM) (Davis, 1985; Davis et al., 1989), technology-task fit model (Goodhue & Thompson, 1995), and decomposed theory of planned behaviour (Adams et al., 1992), among others. Central to this is the TAM (Figure 7.4) developed by Davis (1985). According to TAM, the motivations for adopting technology are the degree of users' perceived usefulness and ease of use. Hence, it is safe to say that this theory puts the power of technology adoption in the adopter's hands and not the inherent benefits of the innovation. According to TAM, the actor's perception of the usefulness and ease of the innovation are the determinants of its adoption or otherwise. Thus, a thorough orientation that informs a good perception will accelerate the diffusion of an innovation. A proliferated understanding or perception is a barrier to BIM adoption.

Presently, the different stakeholders or players in the SACI can have none or low perception of BIM usefulness and ease of use. This assessment is based on the current level of BIM adoption in the SACI, which has been observed to be low (Chimhundu, 2015).

Thus, a change in perception is required to achieve the full adoption of BIM. This requires an orientation of all stakeholders on the usefulness and the BIM ease of use. The efficient change in perception to aid speedy adoption of BIM in the SACI as network relationship is a proven success factor to innovation adoption (Valente, 1996). Another important orientation needed is the orientation about the fitness of BIM for their tasks and the inherent benefits in the adoption of BIM, according to the technology-task fit model (Figures 7.5 and 7.6) (Goodhue & Thompson, 1995).

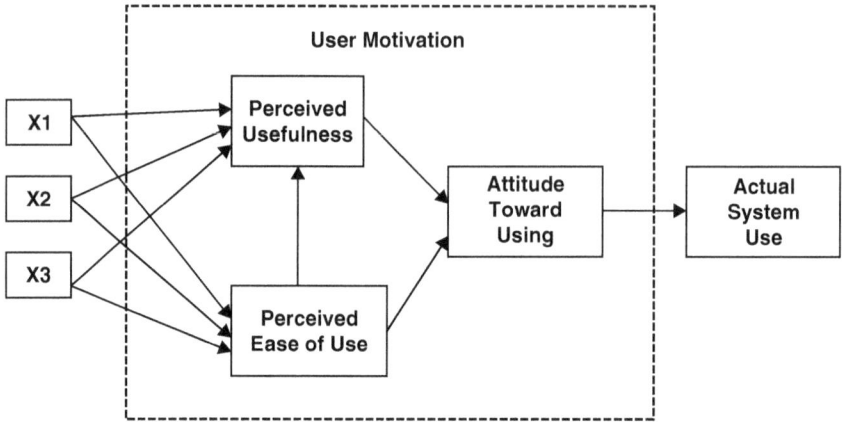

Figure 7.4 Technology acceptance model (Davis, 1985)

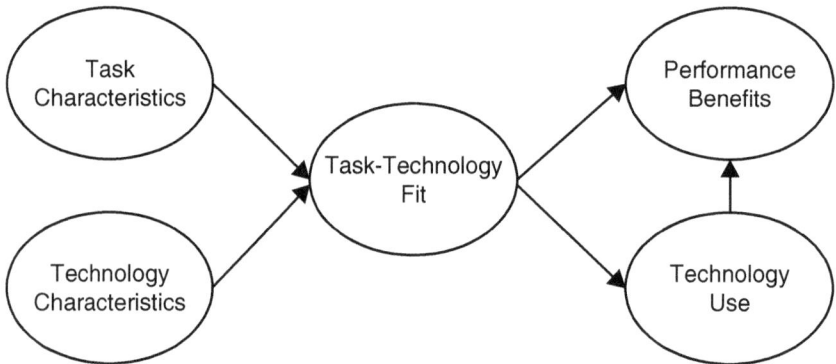

Figure 7.5 Technology task fit model (Goodhue and Thompson, 1995)

Kassem and Succar (2017) identified the stakeholders in the BIM space as policymakers, educational institutions, construction organisations, technology developers, technology service providers, industry associations, communities of practice, and technology advocates.

7.4.4.1 Achieving Stakeholder Orientation

The knowledge and perception of the inherent benefits and demerits are important factors in consumers' decision-making process. Consumers make purchase decisions based on the benefit and returns on investment; cost-benefit analysis is a great determinant of client decision-making. By extension, regarding BIM as a product for the construction industry, the level of information and perception of BIM benefits will determine the BIM maturity level.

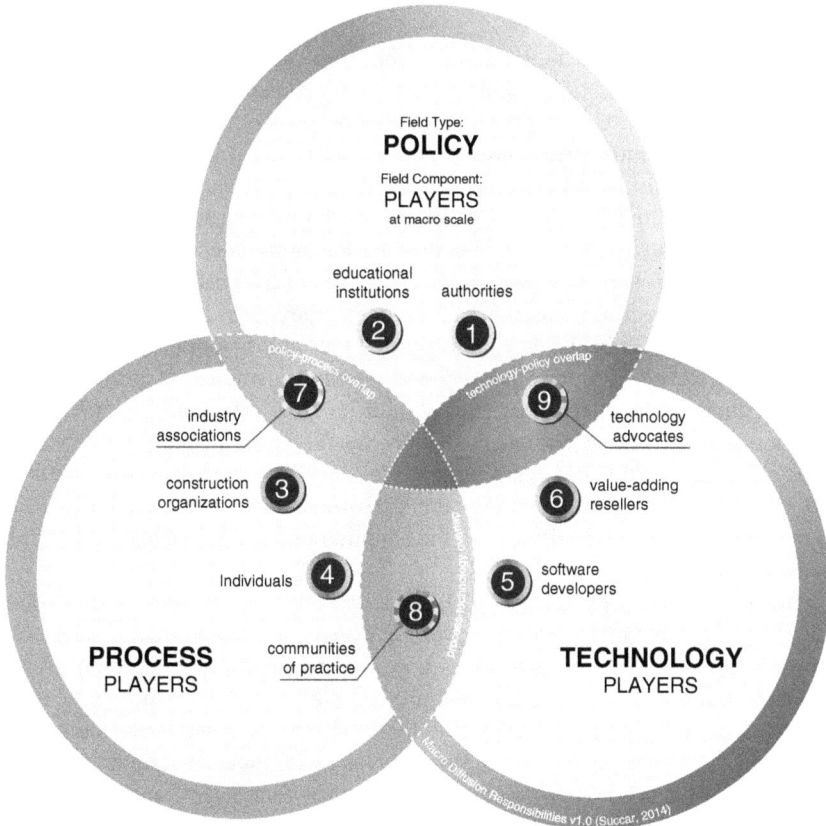

Figure 7.6 Stakeholders in the construction industry (BIM think space)

Stakeholder perceived usefulness of technology has positively influenced technological adoption (Davis, 1985; Louay & Kassem, 2018; Sebastian & Van Berlo, 2010). Adopters are positive to adopt new technology if they have a positive perception as regards the usefulness. On the contrary, if stakeholders fail to be properly educated regarding the BIM benefits, they might be reluctant to adopt. Some of the usefulness of BIM has been established in the literature. For instance, improved ROI, productivity, and increased profit margin (NBS, 2019; Sawhney, 2014) have been observed due to BIM adoption. Contractors in BIM-leading countries have reported fewer errors and omissions, less rework, and lower construction due to investment in BIM (McGraw, 2014). Other stakeholder orientation factors affecting BIM adoption are perceived ease of use (Davis 1985; Sebastian & Van Berlo 2010) and compatibility (Chen, 2013; Rogers, 1962). Despite these documented benefits, BIM adoption has been slow due to a lack of awareness and knowledge (Adekunle et al., 2020; Kekana et al., 2014). Consequently, proper education of stakeholders is required to overcome the lack of BIM awareness and knowledge and provide stakeholders with the correct BIM perception.

7.4.5 Effective BIM Training

The industry landscape is constantly changing due to the introduction of innovations. Complimentarily, skills and BIM educational requirements are changing (Gu et al., 2014). Unlike the United Kingdom, where many universities offer courses in BIM, South African universities are still struggling with teaching BIM in schools. A study by Moodley et al. (2016) on BIM teaching in architecture courses revealed that BIM is taught as a stand-alone course and complimented with design courses. The barrier identified by the study is cost and access to the required software. It is, however, noteworthy that to adequately equip graduates for the fourth industrial revolution requirements, they must be equipped with the required skill set. Gharehbaghi (2015) posited that industry-oriented education is the solution to balancing the industry and learning institutes. It is thus imperative that higher educational institutes should be BIM oriented. The BIM learning tripology depicts an arrangement whereby the learning provider constantly feeds the learner. The BIM learning provider can be higher education institutes and private learning outfits, among others. However, this study will be focusing on the higher education institutes and the adoption of an effective learning method for BIM.

Quality function deployment (QFD) has always been adopted by studies before now to fill this gap. QFD is adopted to improve on an existing product. This method was developed in 1972 in Japan. It is a systematic approach to accommodating customer needs while designing a new product (Jnanesh & Hebbar, 2008). It is a quality assurance method for product delivery that ensures user satisfaction; customer quality or user satisfaction is also referred to as demanded quality (Alptekin & Karsak, 2011). The QFD framework consists of four stages, and each stage's input relies on the previous output.

This framework has been used extensively to develop curriculum in higher education (Alptekin & Karsak, 2011; Upadhyay et al., 2008; Zhang et al., 2016; Zhong et al., 2018). However, for this study, the Delphi process was adopted to effectively capture the industry's requirements as regards BIM training to ensure improvement. However, the delivery method of BIM to learners is missing. The iterative nature of the Delphi process was adopted for this study to fill this gap. The results from the expert's consensus were further validated in this study.

7.4.5.1 Learning Theory

This gap is underpinned by experiential learning theory (Kolb, 1984) and connectivism (Siemens, 2004). These two are learning theories and are thus necessary for the effective training of students. As earlier stated, this gap seeks to determine the effective ways BIM can be taught in learning institutions to students. Thus, it seeks to establish an effective way for BIM knowledge transfer to the students as it relies on technology. Connectivism posits eight principles; these are presented in Table 7.7.

Connectivism rests on the transfer of knowledge and the belief that knowledge resides more than in the human subject. It supports the adoption of technological

Table 7.7 Principles of connectivism

S/N	Connectivism principle
1	Learning and knowledge rests in a diversity of opinions
2	Learning is a process of connecting specialised nodes or information sources
3	Learning may reside in non-human appliances
4	The capacity to know more is more critical than what is currently known
5	Nurturing and maintaining connections is needed to facilitate continual learning
6	The ability to see connections between fields, ideas, and concepts is a core skill
7	Currency (accurate, up-to-date knowledge) is the intent of all connectivist learning activities
8	Decision-making is itself a learning process. Choosing what to learn and the meaning of incoming information is seen through the lens of a shifting reality.

Table 7.8 Learner types

Learner	Preferred instructional method in a formal learning environment.
Diverger	They have a preference for working in groups, listening with an open mind, and receiving personal feedback.
Assimilators	They prefer readings, lectures, exploring analytical models, and thorough thinking things.
Convergers	They prefer experimenting with new ideas, simulations, laboratory assignments, and practical applications. Tend to do well on conventional intelligence tests where there is a single correct answer.
Accommodators	They prefer working with others to get assignments done, setting goals, performing fieldwork, and testing different approaches to completing a project. This category applies the intuitive trial and error method relying on other people for information to solve problems.

tools for the transfer of knowledge. The centrality of this theory lies in the focus on the currency of knowledge as the bedrock of all learning activities. Thus, it supports updating learners' knowledge and updating it to the acceptable industry standard. It, therefore, supports continuous knowledge improvement. Connectivism recognises the dynamism and great changes in the society and thus supports a mode of learning adaptable to these changes (Siemens, 2004).

The experiential learning theory also supports that different learning styles can be combined to achieve effective knowledge transfer. According to Healey and Jenkins (2000), the theory believes in experiential activities like fieldwork and laboratory sessions. Another experiential approach validated in literature is field trips supporting student engagements (Jose et al., 2017). McCarthy (2010) describes it as a holistic, adaptive learning process that combines experience, perception, cognition, and behaviour. This resonates with accommodating the diverse pool of learners as different learners learn using different instructional methods, and this must be considered when designing the effective BIM instructional method. McCarthy (2010) and Massari et al. (2018) posit that there are four types of learners (Table 7.8)

Consequently, to achieve effective BIM training, all the learner types and mode of preferred learning must be considered in designing the teaching method. Also, supporting facilities needed to achieve must be provided.

7.4.5.2 Achieving Effective BIM Training

Achieving BIM maturity is impossible without the learning process tailored to adequately equip learners for the industry dynamism. This is imperative as one of the barriers to BIM adoption is a lack of technical know-how and BIM knowledge (Adekunle et al., 2020; Kekana et al., 2014). To achieve this, the right choice of learning method is essential in teaching the students. This is because BIM is a new technology, and it is being introduced newly into the existing school system.

Different studies have posited that the current academic framework does not align with the industry BIM and that there is a relationship between BIM maturity and the academic framework. These studies posit that a BIM-aligned academic framework is vital for BIM maturity (Moodley et al., 2016; Sotelino et al., 2020; Yilmaz et al., 2019). Also, the choice of teaching strategy and the integration of BIM into the current academic framework are required. Lee et al. (2013) identified some teaching strategies including introducing BIM as a stand-alone course and BIM interactive module, among others. The alignment of the academic framework and the choice of teaching to learners in the tertiary institutions is an essential requisite to achieving BIM maturity.

7.5   The Conceptual Model

7.5.1 People Attribute (PEA)

The human actors are considered very important and central to technological adoption and diffusion. However, this is done at different degrees (Rogers, 1962). The people variable is required to achieve a near-equilibrium in the BIM ecosystem. In achieving technological maturity, BIM inclusive, the people variable has been measured by different studies from different perspectives. This includes but limited to workforce, organisation, industry level, and individual, among others.

This study defines the people's attributes as the human actors involved with implementing the processes and the use of BIM technology in work environments (organisation and project) to achieve BIM maturity in the SACI. For this study, the people variable was studied from three perspectives: the micro, meso, and macro levels. Hence, the study looked at people features from the micro (individual unit) perspective that influence BIM maturity. This was aimed at the different attributes necessary for BIM maturity from the individual professional perspective. At this level, the study identified sub-attributes like competency sets and change resistance, among others.

Also, this study analysed the people attribute from the meso level; this represents the people attributes at the firm or organisation level that influence BIM maturity. This level falls between the micro and the macro levels. The sub-attributes at this level include employee incentives, unstable workforce, and R&D efforts, among others.

The third level of people attribute classification for the study is the macro level. This refers to the people features at the industry level. The sub-attributes studied under this classification include conservative client, government mandate, and the influence of opinion leaders, among others.

7.5.2 Technology Attributes (TEA)

The technological aspect of BIM diffusion has been focused mainly on software, hardware, and network (Succar, 2010a). This implies that most BIM studies analysed the technological aspect, focusing on the software and BIM tools availability and capabilities. This study adopted a more holistic approach by analysing technology in terms of hardware and software requirements, and network integration for BIM usage and maturity in an environment (organisation and project). This basically measures the infrastructure and other attendant factors required for BIM maturity. The software classification studied sub-attributes like interoperability, complementary technologies, spatial capabilities, and innovativeness of software. Another classification focus of the technology construct for the study is the physical infrastructure aspect of technology adoption. This includes sub-attributes such as hardware (availability and upgrade). Lastly, this study considered the data management aspect of technology. It looked at the data exchange influence on BIM maturity, data storage, data richness, and real-time data as attributes that influence BIM maturity. Unlike previous studies, the study adopted a more encompassing view of the BIM maturity of the SACI.

7.5.3 Process Attributes (PSA)

This study adopted the definition of process from the BIM-recognised standard. Processes are every activity carried out by teams during the life cycle of any building asset to achieve effective information management (ISO 19650-2:2018, 2018). Hence, the sub-attributes that were focused on under this attribute were central to the process of BIM adoption. The sub-attributes included process sub-attributes from the firm and project perspectives. The sub-attributes include BIM workflow, appointment protocol, and project complexity, among others.

7.5.4 Stakeholder Orientation Attributes (SOA)

BIM is a collaborative technology; it does involve the adoption by stakeholders to achieve its diffusion. However, stakeholders' orientation forms an important part of BIM maturity; hence, the study is divided it into six focuses. Stakeholder orientation measures the perception and knowledge of the stakeholders in the construction industry as regards BIM. The six classifications of sub-attributes adopted for this study are perceived usefulness, relative advantage, compatibility, trialability, observability, and complexity. All these classifications tested the different parts of stakeholder orientation towards BIM maturity in the SACI.

7.5.5 Ecosystem Regulation Attributes (ERA)

This attribute measures the sets of standards and regulations articulating the alignment of the ecosystem with the BIM requirements. It creates a balance between the structure and the technology. These are a set of regulatory frameworks and their influence on the BIM maturity of the SACI.

7.5.6 Effective BIM Training Attributes (EBT)

This attribute measured the various requirements for successfully integrating BIM into the tertiary institutions as the incubator for industry professionals. This attribute was classified into pedagogical structure, pedagogical strategies, tutor management, assessment methods, and resources. Hence, the study adopted a rather balanced approach to BIM training attributes for BIM maturity.

7.6   Structural Component of Conceptual Model

The conceptualised BIM maturity model is based on the hypothesis that for SACI to achieve BIM maturity, there must be people attributes (PEA), technology attributes (TEC), process attributes (PSA), stakeholder orientation features (SOA), ecosystem regulation attributes (ERA), and effective BIM training attributes (EBT). Thus, it can be said a priori that to achieve BIM3 in SACI, the multidimensional model requires PEA, TEC, PSA, SOA, ERA, and EBT. This conceptualised model shows a relationship between exogenous and variables and other measurements. These variables of measurement, as discussed above, are derived from an extensive review of the literature. The relationship between these variables is represented in Figure 7.7.

7.6.1 Maturity Levels

The BIM maturity shows the extent of BIM capabilities. Also, it has an ordinal scale to capture a gradual and continuous improvement within each BIM capability; it thus provides a performance benchmark. Various BIM maturity models reviewed before now adopted different ordinal scales. For instance, BIM CAREM adopted four levels, BIM3 adopted five levels (like the CMM), and BIM QuickScan adopted six levels. These levels are normally described in an orderly manner whereby the preceding level leads to the next, and it is impossible to jump levels. Maturity levels are well defined, and they establish a level of capability required for BIM maturity (Table 7.9).

Similarly, the developed BIM maturity model in this study adopted a five-level maturity. This aligns with the CMM and the adopted framework by Succar. The five-level maturity was adopted because it provides a systematic and procedural framework for BIM maturity in the SACI. The five maturity levels are defined as follows (Table 7.10):

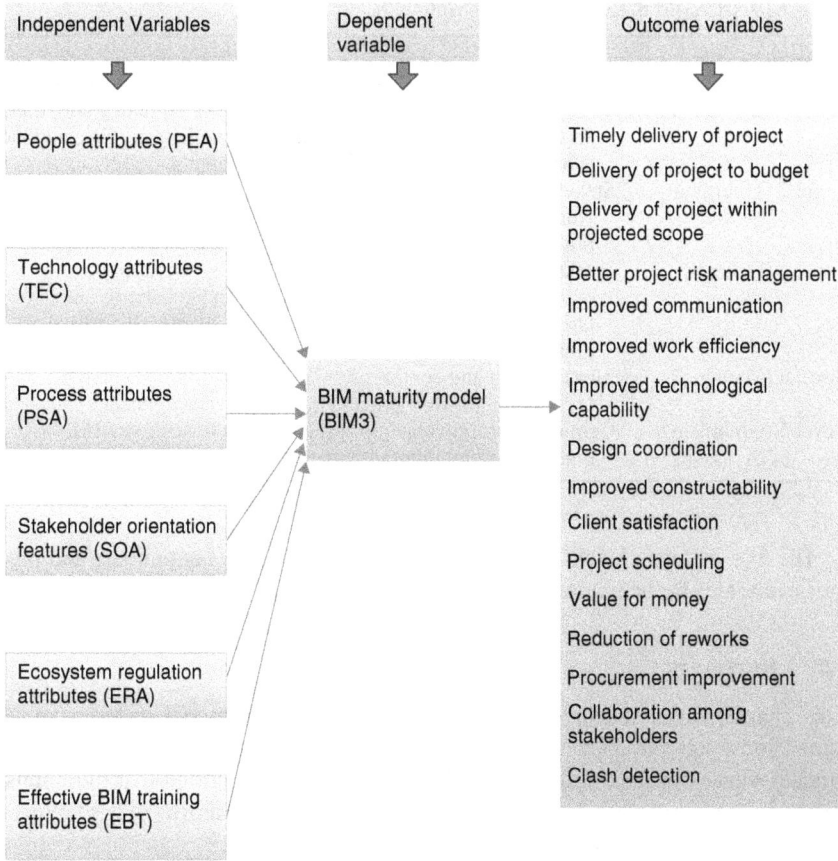

Figure 7.7 BIM maturity conceptual model

Table 7.9 Existing maturity model comparison

Maturity model	BIM CAREM (Yilmaz et al., 2019)	BIM3 (Succar, 2010a)	BIM Quickscan (Berlo et al., 2012)	Organisational BIM assessment profile(Pennsylvania state University, 2013)	VDC BIM scorecard(Kam, Senaratna, Mckinney, Xiao, Song et al., 2013)
Nr of levels	4(0-3)	5	6	6	5
Maturity levels	Incomplete BIM Performed BIM Integrated BIM Optimised BIM	Initial Defined Managed Integrated Optimised	0 to 5	Non-Existent Initial Managed Defined Quantitatively managed Optimising	Conventional Practice Typical Practice Advanced Practice Best Practice Innovative Practice

Table 7.10 BIMMM maturity levels description

Maturity level	Description
Level 1 (Chaotic silos)	The industry lacks control and is reactive to technology adoption. At this level, industry stakeholders are working in silos, and the ecosystem is chaotic regarding BIM adoption.
Level 2 (Emerging)	Stakeholders are still grappling with migrating from the tradition work process to a digitally driven process environment. Stakeholders begin to collaborate and push for BIM adoption.
Level 3 (Defined)	Workflow rework to fit into the collaborative workflow environment to achieve BIM maturity and collaboration. Orientation and education are tailored towards BIM adoption. Stakeholders become reactive.
Level 4 (Managed)	Measurement and control of the process, workflow, client expectation fit work processes, and risk management.
Level 5 (Optimising).	Continuous improvement of process areas to achieve BIM maturity and integration with emerging technologies.

The five maturity levels of this study present systematic, continuous, and progressive scales for achieving BIM maturity in the SACI.

7.7 Summary

This chapter discussed the conceptualised BIM maturity model in terms of its dimensions, sub-variables, and maturity levels. To achieve this, existing maturity models were reviewed, and the identified gaps in them identified that there must be six dimensions to achieve BIM maturity in developing countries. These dimensions are people, process, technology, stakeholder orientation, ecosystem regulation, and effective BIM training. This study adopted five maturity levels: chaotic silos, emerging, defined, managed, and optimising. Achieving each level is a prerequisite to the next. Achieving an optimised maturity will deliver the outcomes as highlighted in the chapter.

References

Adams, D. A., Nelson, R. R., & Todd, P. A. (1992). Perceived Usefulness, Ease of Use, and Usage of Information Technology: A Replication. *MIS Quarterly, 16*(2), 227. https://doi.org/10.2307/249577

Adekunle, S. A., Aigbavboa, C. O., & Ejohwomu, O. A. (2020). BIM Implementation: Articulating the Hurdles in Developing Countries. *8th International Conference on Innovative Production and Construction (IPC)*, 47–54.

Akintola, A., Venkatachalam, S., & Root, D. (2017). New BIM Roles' Legitimacy and Changing Power Dynamics on BIM-Enabled Projects. *Journal of Construction Engineering and Management, 143*(9). https://doi.org/10.1061/(ASCE)CO.1943-7862.0001366

Alptekin, S. E., & Karsak, E. E. (2011). An Integrated Decision Framework for Evaluating and Selecting e-Learning Products. *Applied Soft Computing Journal, 11*(3), 2990–2998. https://doi.org/10.1016/j.asoc.2010.11.023

Arowosegbe, A. A., & Mohamed, S. F. (2015). A Systematic Change Management Capability Maturity Assessment Framework for Contracting Organizations. *American Scientific Journal for Engineering, Technology and Sciences, 13*(1), 88–96.

Azhar, S., Brown, J., & Farooqui, R. (2009). BIM-Based Sustainability Analysis: An Evaluation of Building Performance Analysis Software. *Proceedings of the 45th ASC Annual Conference*. https://pdfs.semanticscholar.org/df0e/5bb056cba01ae8945459e060a1346c992b97.pdf

Barison, M. B., & Santos, E. T. (2011). The Competencies of BIM Specialists: A Comparative Analysis of the Literature Review and Job Ad Descriptions. *Computing in Civil Engineering*. http://www.ascelib

Berlo, L. van, Van Berlo, L., & Hendriks, H. (2012). BIM Quickscan: Benchmark of BIM Performance in the Netherlands. *CIB W78 2012: 29th International Conference*, 17–19. https://www.academia.edu/1905785/BIM_quickscan_benchmark_of_BIM_performance_in_the_netherlands (Accessed: 2 September 2019)

British Standards Institute. (2015). *Bs 1192:2007+a1_2015*.

Carnegie Mellon University. (2005). *Capability Maturity Model ® Integration (CMMI ®) Overview*.

Chan, D. W. M., Olawumi, T. O., & Ho, A. M. L. (2019). Critical Success Factors for Building Information Modelling (BIM) Implementation in Hong Kong. *Engineering, Construction and Architectural Management, 26*(9), 1838–1854. https://doi.org/10.1108/ECAM-05-2018-0204

Chen, Y. (2013). Measurement Models of Building Information Modeling Maturity. In *Purdue University*. Purdue University West Lafayette, Indiana.

Chen, Y., Dib, H., Asce, A. M., Robert Cox, F., Shaurette, M., & Vorvoreanu, M. (2016). Structural Equation Model of Building Information Modeling Maturity. *Journal Construction Engineering Management, 142*(9). https://doi.org/10.1061/(ASCE)CO.1943-7862.0001147

Cheng, J. C. P., & Lu, Q. (2015). A Review of the Efforts and Roles of the Public Sector for BIM Adoption Worldwide. *Journal of Information Technology in Construction, 20*(October), 442–478.

Chimhundu, S. (2015). *A Study on the BIM Adoption Readiness and Possible Mandatory Initiatives for Successful Implementation in South Africa*. University of the Witwatersrand, Johannesburg.

Curtis, B., Hefley, B., & Miller, S. (2009). *People Capability Maturity Model (P-CMM) Version 2.0, Second Edition*. http://www.sei.cmu.edu

Davis, F. D. (1985). A Technology Acceptance Model for Empirically Testing New End-User Information Systems: Theory and Results. In *Management: Vol. Ph.D.* (Issue January 1985). https://doi.org/oclc/56932490

Davis, F. D., Bagozzi, R. P., & Warshaw, P. R. (1989). User Acceptance of Computer Technology: A Comparison of Two Theoretical Models. *Management Science, 35*(8), 982–1003. https://doi.org/10.1287/mnsc.35.8.982

de Bruin, T., Freeze, R., Kulkarni, U., Rosemann, M., Bruin, D., de Bruin, S., Freeze, R., & Carey, W. (2005). Understanding the Main Phases of Developing a Maturity Assessment Model. *16th Australasian Conference on Information Systems, 109*. http://aisel.aisnet.org/acis2005/109

Den Hertog, J. (1975). *Review of Economic Theories of Regulation*. www.uu.nl/rebo/economie/discussionpapers

Drahos, P. (2017). Regulatory Theory. In P. Drahos (Ed.), *Regulatory Theory : Foundations and Applications*. ANU press. https://doi.org/10.1002/9781444320114.ch42

Dremel, C., Overhage, S., Schlauderer, S., & Wulf, J. (2017). Towards a Capability Model for Big Data Analytics. In J. Leimeister & W. Brenner (Eds.), *13th International Conference on Wirtschaftsinformatik, February 12-15, 2017, St. Gallen, Switzerland*, 1141–1155.

Eastman, C. M. (2011). *BIM Handbook : A Guide to Building Information Modeling for Owners, Managers, Designers, Engineers and Contractors*. Wiley. https://books.google.co.za/books?hl=en&lr=&id=aCi7Ozwkoj0C&oi=fnd&pg=PP7&ots=ZbDeOUz7Iq&sig=yvlZ_9-KVeVMEAmjXEog94FzI7Q&redir_esc=y#v=onepage&q&f=false

Eastman, C., Teicholz, P., Sacks, R., & Liston, K. (2008). *BIM Handbook A Guide to Building Information Modeling for Owners, Managers, Designers, Engineers, and Contractors*. John Wiley & Sons, Inc. https://doi.org/10.1093/nq/s7-II.32.110-e

Ezeokoli, F. O., Okolie, K. C., Okoye, P. U., & Belonwu, C. C. (2016). Digital Transformation in the Nigeria Construction Industry: The Professionals' View. *World Journal of Computer Application and Technology*, 4(3), 23–30. https://doi.org/10.13189/wjcat.2016.040301

Ezeokoli, F., Okoye, P., & Nkeleme, E. (2016). Factors Affecting the Adaptability of Building Information Modelling (BIM) for Construction Projects in Anambra State Nigeria. *Journal of Scientific Research and Reports*, 11(5), 1–10. https://doi.org/10.9734/JSRR/2016/26563

Geels, F. W. (2004). From Sectoral Systems of Innovation to Socio-Technical Systems: Insights about Dynamics and Change from Sociology and Institutional Theory. *Research Policy*, 33(6–7), 897–920. https://doi.org/10.1016/j.respol.2004.01.015

Gharehbaghi, K. (2015). *The Importance of Industry Links in Teaching Pedagogy : A Higher Education*. 5(1), 17–23.

Giddens, A. (1984). *The Constitution of Society Outline of the Theory of Structuration*. University of California Press.

Goodhue, D. L., & Thompson, R. L. (1995). Task-Technology Fit and Individual Performance. *MIS Quarterly*, 19(2),213–236.

Gu, N., & London, K. (2010). Understanding and Facilitating BIM Adoption in the AEC Industry. *Automation in Construction*, 19, 988–999. https://doi.org/10.1016/j.autcon.2010.09.002

Gu, N., Singh, V., & London, K. (2014). BIM Ecosystem : The Coevolution of Products , Processes , and People. In K. M. Kensek & D. Noble (Eds.), *Building Information Modeling: BIM in Current and Future Practice* (Issue 1, pp. 197–210). John Wiley & Sons, Inc. https://doi.org/10.1002/9781119174752.ch15

Healey, M., & Jenkins, A. (2000). Kolb's Experiential Learning Theory and Its Application in Geography in Higher Education. *Journal of Geography*, 99(5), 185–195. https://doi.org/10.1080/00221340008978967

ISO 19650-1:2018. (2018). *Organization and Digitization of Information about Buildings and Civil Engineering Works, Including Building Information Modelling (BIM) - Information Management Using Building Information Modelling*, (19650–1). ISO.

ISO 19650-2:2018. (2018). *Organization and Digitization of Information about Buildings and Civil Engineering Works, Including Building Information Modelling (BIM) - Information Management Using Building Information Modelling - Part 2: Delivery Phase of the Assets (ISO 19650-2:2018.*

ISO 19650-5:2020. (2020). *Organization and Digitization of Information about Buildings and Civil Engineering Works, Including Building Information Modelling (BIM) - Information Management Using Building Information Modelling - Part 5: Security-Minded Approach to Information Manag.*

Jnanesh, N. A., & Hebbar, K. C. (2008). Use of Quality Function Deployment Analysis in Curriculum Development of Engineering Education and Models for Curriculum Design and Delivery. *Proceedings of the World Congress on Engineering and Computer Science 2008 WCECS 2008, October 22 – 24, 2008, San Francisco, USA*, 22–25.

Joblot, L., Paviot, T., Deneux, D., & Lamouri, S. (2019). Building Information Maturity Model Specific to the Renovation Sector. *Automation in Construction*, *101*, 140–159. https://doi.org/10.1016/j.autcon.2019.01.019

Jones, M., & Karsten, H. (2003). *Review: Structuration Theory and Information Systems Research* (Research Papers in Management Studies). www.jims.cam.ac.uk

Jose, S., Patrick, P. G., & Moseley, C. (2017). Experiential Learning Theory: The Importance of Outdoor Classrooms in Environmental Education. *International Journal of Science Education, Part B*, *7*(3), 269–284. https://doi.org/10.1080/21548455.2016.1272144

Kam, C., Senaratna, D., Mckinney, B., Xiao, Y., & Song, M. (2013). *The VDC Scorecard: Formulation and Validation*. Center for Integrated Facility Engineering: Stanford University.

Kam, C., Senaratna, D., Mckinney, B., Xiao, Y., Song, M., & Mckinney, B. (2013). *The VDC Scorecard: Evaluation of AEC Projects and Industry Trends*. https://stacks.stanford.edu/file/druid:st437wr3978/WP136.pdf

Kassem, M., Succar, B., & Dawood, N. (2015). Building Information Modeling: Analyzing Noteworthy Publications of Eight Countries Using a Knowledge Content Taxonomy. In *Building Information Modeling: Applications and Practices* (Issue 61, pp. 329–371). https://doi.org/10.1061/9780784413982.ch13

Kassem, M., Liyana, N., Raoff, A., & Ouahrani, D. (2018). Identifying and Analyzing BIM Specialist Roles using a Competency-Based Approach. In M. J. Skibniewski & M. Hajdu (Eds.), *Proceedings of the Creative Construction Conference*. https://doi.org/10.3311/CCC2018-135

Kassem, M., & Succar, B. (2017). *Macro BIM Adoption: Comparative Market Analysis*. *81*(September 2016), 286–299. https://doi.org/10.1016/j.autcon.2017.04.005

Kekana, T., Aigbavboa, C., & Thwala, W. (2014). Building Information Modelling (BIM): Barriers in Adoption and Implementation Strategies in the South Africa Construction Industry. *International Conference on Emerging Trends in Computer and Image Processing (ICETCIP'2014) Dec. 15-16, 2014 Pattaya (Thailand)*.

Khoshfetrat, R., Sarvari, H., Chan, D. W. M., & Rakhshanifar, M. (2020). Critical Risk Factors for Implementing Building Information Modelling (BIM): A Delphi-Based Survey. *International Journal of Construction Management*. https://doi.org/10.1080/15623599.2020.1788759

Kolb, D. (1984). *Experiential Learning: Experience as the Source of Learning and Development* (1st ed.). Prentice Hall. www.learningfromexperience.com/images/uploads/process-of-experiential-learning.pdf

Latham, M. (1994). Constucting the Team. In *Joint Review of Procurement and Contractual Arrangements in the United Kingdom Construction Industry*. https://doi.org/10.1017/CBO9781107415324.004

Lee, N., Dossick, C. S., & Foley, S. P. (2013). Guideline for Building Information Modeling in Construction Engineering and Management Education. *Journal of Professional Issues in Engineering Education Practice*, *4*(139), 266–274. https://doi.org/10.1061/(ASCE)EI.1943-5541.0000163

Liang, C., Lu, W., Rowlinson, S., & Zhang, X. (2016). Development of a Multifunctional BIM Maturity Model. *Journal of Construction Engineering and Management*, *142*(11). https://doi.org/10.1061/(ASCE)CO.1943-7862.0001186

Louay, A., & Kassem, M. (2018). A Unified BIM Adoption Taxonomy: Conceptual Development , Empirical Validation and Application. *Automation in Construction, 96*(November 2017), 103–127. https://doi.org/10.1016/j.autcon.2018.08.017

Massari, G.-A. F.-M. M., Kamantauskiene, V., Zeynep Alat, C., Mesquita, M. T., Verheij, J. K., & Zirina, T. (2018). *A Handbook on Experiential Education: Pedagogical Guidelines for Teachers and Parents* (G.-A. F.-M. M. Massari, V. Kamantauskiene, C. Zeynep Alat, M. T. Mesquita, J. K. Verheij, & T. Zirina (Eds.)). Editura Universităţii, Alexandru Ioan Cuza. www.editura.uaic

Mayo, G., Wu, W., McCuen, T., Issa, R. R. A., & Smith, D. (2018). Implementation of the BIM Body of Knowledge (BOK) Framework for Program Planning in Academia. *12th BIM Academic Symposium & Job Task Analysis Review, 102*(c), 1–90.

McCarthy, M. (2010). Experiential Learning Theory: From Theory To Practice. *Journal of Business & Economics Research (JBER), 14*(3), 91–100. https://doi.org/10.19030/jber.v14i3.9749

McGee, R., & Warms, R. (2013). Network Theory/Social Network Analysis. In R. J. M. & R. L. Warms (Eds.), *Theory in Social and Cultural Anthropology: An Encyclopedia* (Issue August 2016). SAGE Publications, Inc. https://doi.org/10.4135/9781452276311.n196

McGraw Hill. (2014). *The Business Value of BIM for Construction in Major Global Markets: How Contractors Around the World Are Driving Innovation With Building Information Modeling.* www.construction.com

Moodley, V., Mathye, K., & Radebe, S. (2016). *Teaching BIM in Schools of Architecture of South African Universities.* University of the Witwatersrand, South Africa.

Moore, J. E. (1996). *The death of competition: leadership and strategy in the age of business ecosystems.* Harper Business.

Moosa, I. A. (2015). Good Regulation, Bad Regulation: The Anatomy of Financial Regulation. In P. Molyneux (Ed.), *Palgrave Macmillan Studies in Banking and Financial Institutions* (1st ed.). Springer.

Munir, M., Kiviniemi, A., Jones, S. W., Finnegan, S., & Mêda, P. (2019). Development of a BIM Asset Maturity Model. In *Advances in ICT in Design, Construction and Management in Architecture, Engineering, Construction and Operations (AECO): Proceedings of the 36th CIB W78 2019 Conference.* University of Northumbria, 360–368.

NBS. (2019). National BIM Report 2019. *National BIM Report 2019 The Definitive Industry Update*, 1–28. https://doi.org/10.1017/CBO9781107415324.004

Nelson, R. R., & Winter, S. G. (1983). An Evolutionary Theory of Economic Change. In *The Economic Journal, 93*(371). https://doi.org/10.2307/2232409

Pennsylvania state University. (2013). BIM Planning Guide for Facility Owners. In *The Pennsylvania State University, University Park, PA, USA: Vol. Version 2.* http://bim.psu.edu

Peter, M., & Grivas, S. G. (2016). The Need of a Framework for the Digital Transformation of Industry Ecosystems Handling Intercompany Collaborative Workflows. *In COLLA16– The Sixth International Conference on Advanced Collaborative Networks, Systems and Applications, November 13-17, 2016, Barcelona, Spain. IARIA XPS Press.*

Rogers, E. M. (1962). *Diffusion of Innovations* (Third Edit). The Free press.

Rose, J., & Scheepers, R. (2001). Structuration Theory And Information System Development-Frameworks For Practice. *Global Co-Operation in the New Millennium The 9th European Conference on Information Systems Bled, Slovenia, June 27-29, 2001*, 217–231.

Sacks, R., Girolami, M., & Brilakis, I. (2020). Building Information Modelling, Artificial Intelligence and Construction Tech. *Developments in the Built Environment*, 100011. https://doi.org/10.1016/j.dibe.2020.100011

Saka, A. B., & Chan, D. W. (2019). Knowledge, Skills and Functionalities Requirements for Quantity Surveyors in Building Information Modelling (BIM) Work Environment: An International Delphi Study. *Architectural Engineering and Design Management*, *16*(3), 227–246. https://doi.org/10.1080/17452007.2019.1651247

Sawhney, A. (2014). *State of BIM Adoption and Outlook in India.* http://www.fig.net/ resources/proceedings/fig_proceedings/fig2014/ppt/ss36/ss36_kavanagh_7434.pdf

Scaringella, L., & Radziwon, A. (2018). Innovation, Entrepreneurial, Knowledge, and Business Ecosystems: Old Wine in New Bottles? *Technological Forecasting and Social Change*, *136*(December 2015), 59–87. https://doi.org/10.1016/j.techfore.2017.09.023

Schumacher, A., Erol, S., & Sihn, W. (2016). A Maturity Model for Assessing Industry 4.0 Readiness and Maturity of Manufacturing Enterprises. *Procedia CIRP*, *52*, 161–166. https://doi.org/10.1016/j.procir.2016.07.040

Sebastian, R., & Van Berlo, L. (2010). Tool for Benchmarking BIM Performance of Design, Engineering and Construction Firms in the Netherlands. *Architectural Engineering and Design Management*, *6*(SPECIAL ISSUE), 254–263. https://doi.org/10.3763/aedm.2010. IDDS3

Siemens, G. (2004). *Connectivism as a Learning Theory for the Digital Age* (pp. 14–16). elearnspace.org.

Smith, A., & Raven, R. (2012). What Is Protective Space? Reconsidering Niches in Transitions to Sustainability. *Research Policy*, *41*(6), 1025–1036. https://doi.org/10.1016/ j.respol.2011.12.012

Sotelino, E. D., Natividade, V., Saad, C., & Do Carmo, T. (2020). Teaching BIM and Its Impact on Young Professionals. *Journal of Civil Education.* https://doi.org/10.1061/ (ASCE)EI.2643-9115.0000019

Succar, B. (2009). Building Information Modelling Framework: A Research and Delivery Foundation for Industry Stakeholders. *Automation in Construction*, *18*(3), 357–375. https://doi.org/10.1016/j.autcon.2008.10.003

Succar, B. (2010a). Building Information Modelling Maturity Matrix. In *Handbook of Research on Building Information Modelling and Construction and Construction Infomatics: Concepts and Technologies* (pp. 65–103). https://doi.org/10.4018/978-1-60566-928-1.ch004

Succar, B. (2010b). The Five Components of BIM Performance Measurement. *Proceedings of CIB World Congress, Salford.* https://doi.org/10.1136/bmj.3.5560.312-a

Succar, Bilal & Agar, Carl & Beazley, Scott & Berkemeier, Paul & Choy, Richard & Giangregorio, Rosetta & Donaghey, Steven & Linning, Chris & Macdonald, Jennifer & Perey, Rodd & Plume, Jim. (2012). BIM Education, BIM in Practice. Australian Institute of Architects.

Succar, B., Sher, W., & Williams, A. (2012). Measuring BIM Performance : Five Metrics. *Architectural Engineering and Design Management*, *8*(2), 120–142. https://doi.org/10. 1080/17452007.2012.659506

Succar, B., Sher, W., & Williams, A. (2013). An Integrated Approach to BIM Competency Assessment, Acquisition and Application. *Automation in Construction*, *35*, 174–189. https://doi.org/10.1016/j.autcon.2013.05.016

Teece, D. J. (2017). Towards a Capability Theory of (Innovating) Firms : Implications for Management and Policy. *Cambridge Journal of Economics*, *41*(April), 693–720. https:// doi.org/10.1093/cje/bew063

Teece, D. J. (2019). A Capability Theory of the Firm : An Economics and (Strategic) Management Perspective Management Perspective. *New Zealand Economic Papers*, *0*(0), 1–43. https://doi.org/10.1080/00779954.2017.1371208

Teece, D. J., Pisano, G., & Shuen, A. (1997). Dynamic Capabilities and Strategic Management. *Dynamic Capabilities and Strategic Management, 18*(7), 27–52. https://doi.org/10.1142/9789812834478_0002

Troiani, E., Mahamadu, A.-M., Manu, P., Kissi, E., Aigbavboa, C., & Oti, A. (2020). Macro-maturity factors and their influence on micro-level BIM implementation within design firms in Italy. *Architectural Engineering and Design Management*. https://doi.org/10.1080/17452007.2020.1738994

Upadhyay, R. K., Gaur, S. K., & Agrawal, V. P. (2008). Model Development Using Fuzzy Quality Function Deployment (FQFD) to Assess Student Requirement in Engineering Institutions: An Indian Prospective. *Certified International Journal of Engineering Science and Innovative Technology (IJESIT), 2*(5), 310–317.

Valente, T. W. (1996). Social network thresholds in the diffusion of innovations. *Social Networks, 18*, 60–89.

Walrave, B., Talmar, M., Podoynitsyna, K. S., Romme, A. G. L., & Verbong, G. P. J. (2018). A multi-level perspective on innovation ecosystems for path-breaking innovation. *Technological Forecasting and Social Change, 136*(December 2016), 103–113. https://doi.org/10.1016/j.techfore.2017.04.011

WEF, & BCG. (2016). *Shaping the Future of Construction A Breakthrough in Mindset and Technology*. World Economic Forum.

Wu, C., Xu, B., Mao, C., & Li, X. (2017). Overview of BIM Maturity Measurement Tools. *Journal of Information Technology in Construction (ITcon), 22*, 35. http://www.itcon.org/2017/3

Yilmaz, G. (2017). *BIM-CAREM: A Reference Model for Building Information Modelling Capability Assessment*. The Middle East Technical University.

Yilmaz, G., Akcamete, A., & Demirors, O. (2019). A Reference Model for BIM Capability Assessments. *Automation in Construction, 101*(January 2018), 245–263. https://doi.org/10.1016/j.autcon.2018.10.022

Zhang, J., Schmidt, K., & Li, H. (2016). BIM and sustainability Education: Incorporating Instructional Needs into Curriculum Planning in CEM Programs Accredited by ACCE. *Sustainability (Switzerland), 8*(6), 1–32. https://doi.org/10.3390/su8060525

Zhong, J., Guo, A., Fu, Y., & Wang, H. (2018). Application of QFD in China's Higher Education: A Bibliometric Study. *4th International Conference on Economics, Social Science, Arts, Education and Management Engineering (ESSAEME 2018) Application, 204*(Essaeme), 249–254. https://doi.org/10.2991/essaeme-18.2018.47

Appendix

Sub-attributes under each construct

Constructs	Variables	Reference
People	Training	(Chen, 2013; Sebastian & Van Berlo, 2010; Succar, 2010a; Succar et al., 2013; Yilmaz et al., 2019)
	Skills and expertise	(Chen, 2013; Schumacher et al., 2016; Succar, 2009, 2010b; Succar, Sher et al., 2012; Succar et al., 2013; Yilmaz et al., 2019)
	Leadership	(Arowosegbe & Mohamed, 2015; Chen, 2013; Schumacher et al., 2016; Succar, 2009; Succar, Sher et al., 2012; Yilmaz et al., 2019)
	Culture	(Berlo et al., 2012; de Bruin et al., 2005; Schumacher et al., 2016; Sebastian & Van Berlo, 2010; Succar, Sher, et al., 2012; Yilmaz et al., 2019)
	Management framework	(Berlo et al., 2012; Kam, Senaratna, Mckinney, Xiao, Song, et al., 2013; Liang et al., 2016; Louay & Kassem, 2018; Succar, Sher, et al., 2012; Yilmaz et al., 2019)
	Strategic alignment	(Chen, 2013; de Bruin et al., 2005; Kam, Senaratna, Mckinney, Xiao, & Song, 2013; Munir et al., 2019; Sebastian & Van Berlo, 2010; Succar, 2010a; Succar, Sher, et al., 2012; Teece et al., 1997; Teece, 2019; Yilmaz et al., 2019)
	Financial resources	(Kam, Senaratna, Mckinney, Xiao, & Song, 2013; Louay & Kassem, 2018; Sebastian & Van Berlo, 2010)
	Organisational structure	(Louay & Kassem, 2018; Succar, 2010a)
	Competency sets	(Barison & Santos, 2011; Mohamad Kassem et al., 2018)
	Actor role definition	(Akintola et al., 2017)

(*Continued*)

Constructs	*Variables*	*Reference*
	Actor role qualification	(Akintola et al., 2017)
	Resistance to change	(Chimhundu, 2015)
	Willingness to share information	(Khoshfetrat et al., 2020)
	Recruitment challenge	(Khoshfetrat et al., 2020)
	Management support	(Khoshfetrat et al., 2020; Teece et al., 1997)
	Unequal adoption readiness	(Khoshfetrat et al., 2020)
	Individual/firm awareness level	(Chan et al., 2019)
	Other stakeholders' BIM involvement	(Chan et al., 2019)
	Client support and involvement	(Chan et al., 2019)
	Information management within organisation	(Chan et al., 2019)
	Aligning and managing resources and teams to focus on digital services	(Ezeokoli, Okolie, Okoye, & Belonwu, 2016)
	Technological mindset	(Ezeokoli, Okolie, Okoye, & Belonwu, 2016)
	Employee incentive	(Ezeokoli et al., 2016)
	Unstable workforce	(WEF & BCG, 2016)
	Conservative clients	(WEF & BCG, 2016)
	Multiple stakeholders with diverse interests/need	(WEF & BCG, 2016)
Technology	Applications	(Berlo et al., 2012; Chen, 2013)
	Hardware	(Chen, 2013; Kam, Senaratna, Mckinney, Xiao, & Song, 2013; Liang et al., 2016; Schumacher et al., 2016; Succar, 2009, 2010b; Succar, Sher, et al., 2012)
	Innovation	(Chen, 2013; Succar, Sher, et al., 2012)
	Data security	(ISO 19650-5:2020, 2020; Liang et al., 2016; Saka & Chan, 2019; Succar, 2010b; Yilmaz et al., 2019)
	Software	(Kam, Senaratna, Mckinney, Xiao, & Song, 2013; Succar, 2009, 2010b; Succar, Sher, et al., 2012; Yilmaz et al., 2019)
	Interoperability	(Chen, 2013; Kam, Senaratna, Mckinney, Xiao, & Song, 2013; Liang et al., 2016; Louay & Kassem, 2018; Saka & Chan, 2019)
	Complimentary technologies	(Sacks et al., 2020; Succar, 2010b)
	Database technology	(Succar, 2010b)
	Data exchange	(Succar, 2010b, 2010a)
	Data storage	(Succar, 2010b)
	Infrastructure	(Succar, 2009; Succar, Sher, et al., 2012)

(Continued)

Constructs	Variables	Reference
	Intellectual property	(Khoshfetrat et al., 2020)
	Cost of software renewal	(Khoshfetrat et al., 2020)
	Cyber security	(Khoshfetrat et al., 2020)
	Availability/affordability of cloud-based technology	(Khoshfetrat et al., 2020)
	Technical support by software vendors	(Khoshfetrat et al., 2020)
Process	Information management	(Berlo et al., 2012; ISO 19650-1:2018, 2018; ISO 19650-2:2018, 2018; Joblot et al., 2019; Sebastian & Van Berlo, 2010; Succar, Sher, et al., 2012; Yilmaz et al., 2019)
	Standardisation	(Arowosegbe & Mohamed, 2015; ISO 19650-2:2018, 2018; Liang et al., 2016; Schumacher et al., 2016; Sebastian & Van Berlo, 2010)
	Quality control	(Chen, 2013; Liang et al., 2016; Yilmaz et al., 2019)
	Roles and responsibilities	(Chen, 2013; ISO 19650-1:2018, 2018; ISO 19650-2:2018, 2018; Liang et al., 2016)
	Communication and collaboration	(ISO 19650-1:2018, 2018; Louay & Kassem, 2018)
	Stakeholders management	(Kam, Senaratna, Mckinney, Xiao, & Song, 2013)
	Research and development	(Succar, 2009, 2010a, 2010b; Teece, 2017; Yilmaz et al., 2019
	Information sharing	(ISO 19650-5:2020, 2020; Liang et al., 2016; Yilmaz et al., 2019)
	Data accuracy	(Chen, 2013; Liang et al., 2016; Yilmaz et al., 2019)
	Level of detail	(Chen, 2013; Kam, Senaratna, Mckinney, Xiao, & Song, 2013)
	Workflow	(Chen et al., 2016; ISO 19650-1:2018, 2018)
	Execution plan	(ISO 19650-1:2018, 2018; ISO 19650-2:2018, 2018)
	Information delivery cycle	(ISO 19650-1:2018, 2018)
	Information requirements	(ISO 19650-1:2018, 2018)
	Information protocol	(ISO 19650-2:2018, 2018)
	Appointment and mobilisation protocols	(ISO 19650-2:2018, 2018)
	Change management	(Chen et al., 2016)
	Information value assessment	(Dremel et al., 2017)
	Risk management	(Azhar et al., 2009)
	Project complexity	(Chan et al., 2019)
	Early involvement of project team	(Gu & London, 2010)

(*Continued*)

Constructs	Variables	Reference
Stakeholder orientation	Perceived usefulness	(Davis, 1985; Louay & Kassem, 2018; Sebastian & Van Berlo, 2010)
	Perceived ease of use	(Davis, 1985; Louay & Kassem, 2018; Sebastian & Van Berlo, 2010)
	Relative advantage	(Louay & Kassem, 2018; Rogers, 1962)
	Compatability	(Chen, 2013; Louay & Kassem, 2018; Rogers, 1962)
	Trialability	(Louay & Kassem, 2018; Rogers, 1962)
	Interpersonal relationship/ network	(Louay & Kassem, 2018; Valente, 1996)
	Performance benefits	(Goodhue & Thompson, 1995)
	Opinion leader	(Gokcen Yilmaz et al., 2019)
	Attitude	(Davis, 1985; Kam, Senaratna, Mckinney, Xiao, & Song, 2013)
Effective BIM training	Academic framework	(Mayo et al., 2018; Moodley et al., 2016; Sebastian & Van Berlo, 2010; Sotelino et al., 2020; Succar, 2010b; Succar, Agar, et al., 2012; Succar, Sher et al., 2012; Yilmaz et al., 2019)
	Pedagogy standards and benchmarks	(Mayo et al., 2018; Succar, Agar, et al., 2012; Succar, Sher, et al., 2012; Yilmaz et al., 2019)
	Infrastructure	(Succar, 2009; Succar, Sher, et al., 2012)
	Hardware	(Chen, 2013; Kam, Senaratna, Mckinney, Xiao, & Song, 2013; Liang et al., 2016; Schumacher et al., 2016; Succar, 2009, 2010b; Succar, Sher, et al., 2012)
	Software	(Kam, Senaratna, Mckinney, Xiao, & Song, 2013; Succar, 2009, 2010b; Succar, Sher, et al., 2012; Yilmaz et al., 2019)
	Interactive modules	(Lee et al., 2013)
	Cross-curriculum teaching module	(Lee et al., 2013)
	Stand-alone course	(Lee et al., 2013)
	BIM-based capstone project experience	(Lee et al., 2013)
Ecosystem regulation	Government mandate	(Louay & Kassem, 2018; Kassem et al., 2015; Succar, 2010a)
	Regulation and industry standards	(Louay & Kassem, 2018; Succar, 2009; Yilmaz et al., 2019)
	Other stakeholder regulations and guidelines	(ISO 19650-5:2020, 2020; Louay & Kassem, 2018; Kassem et al., 2015; Succar, 2009; Succar, Sher, et al., 2012)
	Delivery benchmarks	(Kam, Senaratna, Mckinney, Xiao, & Song, 2013; Kassem et al., 2015; Succar, 2010a; Succar, Sher, et al., 2012; Troiani et al., 2020; Yilmaz et al., 2019)
	Responsibilities, rewards	(Chen, 2013; Succar, 2009, 2010a; Succar, Sher, et al., 2012)
	Third-party management	(ISO 19650-5:2020, 2020)
	Legal framework	(Chan et al., 2019)

8 Evaluation of the BIM Maturity Model in a Developing Country

8.1 Introduction

This chapter explores the conceptualised building information modelling (BIM) maturity model for developing countries using the South African construction industry (SACI) as a laboratory case. This chapter explores the dimensions, sub-variables, and maturity levels conceptualised in the previous chapter. To achieve this, a Delphi approach was adopted. This chapter provides an overview of the Delphi method, and how the study adopted it was also explained in detail. Furthermore, this chapter presents how the results were analysed and interpreted.

The Delphi adopted an iterative process consisting of two rounds of Delphi survey before achieving the expert consensus. In this chapter, the results of the Delphi rounds and the computations are also presented. The computation for the influence of each of the attributes is presented. The background information and composition of the panel and the general discussion of the Delphi results were also presented in this chapter.

8.2 The Delphi Method

Delphi is a systematic method for solicitation and information collection on a topic of interest (Turoff, 1970). It is a technique employed for collecting judgement to overcome the weakness of eliciting a response from one expert, a one-shot group average round-table discussion (Clayton, 1997). Linstone and Turoff (2002) describe it as the structuring of human communications. It is one of the most commonly used group interview approaches; others are focus groups and brainstorming (Macphail, 2001). However, Delphi provides the most reliable consensus of opinion from a group (Clayton, 1997). The respondents for Delphi are called experts, and it is different from other techniques because it relies on expert consensus. Also, it does not require the close proximity of experts, and it is iterative. Furthermore, it cuts down on cost and time (Yousuf, 2007).

The application of the Delphi method is dated back to the 1940s by the RAND Corporation in the forecasting of technology. According to the first publication on this technique by Helmer and Rescher (1960), this technique is permissible in fields where scientific instruments have not been fully developed; hence, experts'

DOI: 10.1201/9781003373919-10

testimony and opinions are permissible for forecasting. However, it should be noted that Delphi is older than the RAND application. It is dated back to an ancient Greek myth called the "Delphic Oracle", which birthed the term "Delphi". In this old Greek mythology, it holds that a "chosen one" possesses the ability to predict the future accurately. According to Turoff (1970), it is carried out to explore various assumptions leading to different judgements, and this information is generated through the respondent's consensus judgement. The respondents for a Delphi are referred to as experts. There are different Delphi types, including conventional, real-time, policy, argument, and dissaggregative Delphi (Milevska-kostova & Dunn, 2014; Turoff, 1970). Also, the Delphi method can be conducted either in the conventional way known as the "paper-and-pencil" with a version overseen by a human team and the Delphi conferences coordinated by a programmed computer (Linstone & Turoff, 2002).

Some of the Delphi technique characteristics identified in the literature (Milevska-kostova & Dunn, 2014) include the geographical dispersal of experts. Also, it allows for a systematic approach to a complex problem by experts. It also allows for at least an opportunity to reevaluate original answers by experts. A unique feature of the Delphi technique is that it presents a single aggregated response and not individual expert opinions (NSA, 2011); hence every expert opinion contributes to the final result. Although the Delphi technique possesses some advantages, it relies on certain principles guiding its execution. Some of the principles guiding the Delphi technique that makes it distinguished (NSA, 2011) are:

- Anonymity: This principle entails that Delphi experts are not aware of each other and individuals responses. This allows for the truthful contribution of experts hence allowing for equality among experts. Also, it discourages the dominance of opinion and enables all experts to contribute freely. Thus, it ensures objectivity by eliminating the bandwagon effect.
- Iteration: The Delphi method is an iterative process, i.e., it allows for repetitiveness. This is done to achieve consensus within experts through refinement of their contribution. The iterative process does not imply that all experts must agree, but results must reflect the consensus in experts' positions. It requires expert consensus or stakeholder disagreement (Milevska-kostova & Dunn, 2014) for the rounds of a Delphi to be put to an end. Expert consensus is reached when after several rounds, the expert's opinions appear to converge. The rigorous process of achieving consensus within experts can lead to fatigue among them, hence the use of stakeholder disagreement. Stakeholder disagreement involves generating views on a major issue especially opposing views.
- Controlled feedback: The Delphi process is managed, and feedback is synthesised to determine consensus. Between rounds, the researcher collates, analyses and contact experts to achieve consensus. This requires the meticulous ability of the researcher and also analytical skills.
- Selection of experts: This is critical as the Delphi survey results are dependent on the aggregated opinions of the constituent experts for the study. An expert

has been defined as someone who has vast knowledge about a field study. Referring to someone as an expert for a particular study depends on the study's aims as it is not a most they are professionals in that field (McMillan et al., 2016). Furthermore, panel choice should be heterogeneous (Linstone & Turoff, 2002). The researcher must recruit experts who will be involved in the study; it is thus advisable to go for experts who will be affected by the study outcome (Hasson et al., 2000). Also, impartial experts are preferred and encouraged.

The number of rounds for a Delphi differs as the aim is to achieve consensus among experts. Consensus is determined by the sum of judgements, consistency of opinions between rounds or a subjective level of central tendencies (Holey et al., 2007a). However, the popular number of rounds is between two and three rounds (De Loë et al., 2016), but there have been studies having more than three rounds (Chan & Chan, 2012).

Despite the peculiar characteristics of the Delphi method, it has its weaknesses. According to NSA (2011), the Delphi technique has three major weaknesses. Firstly, the absence of the right types of experts in a field will not produce the right forecast. Secondly, experts' responses are mostly not supported by facts. Hence facts are not considered. It thus means that widely held misconceptions might be generalised. Thirdly the technique applies a simplified approach to a complex problem that might not apply in all situations.

Over the years, the Delphi method has gained popularity and adoption in several fields. These fields include medical science (Eubank et al., 2016; Holey et al., 2007b; Tetzlaff et al., 2012; Woodcock et al., 2020), logistics management (Gossler et al., 2019), bioeconomy (Devaney & Henchion, 2018), IT and IS (Okoli & Pawlowski, 2004), construction (Khoshfetrat et al., 2020).

For this study, the Delphi method was also adopted because the study is predictive. It predicts the BIM maturity factors for the SACI. Also, it requires subjective experts' opinions to understand the complex nature of BIM diffusion in the SACI. Also, the experts required for the study cannot be brought together physically (Yousuf, 2007). This is particularly useful in the period of the coronavirus pandemic where social distancing and lockdown measures are in place. In reporting the Delphi process, the study adopted the outline by Hasson et al. (2000). The study provides the outline for reporting the Delphi method. This includes stating the research problem, articulating the research rationale, the methodology, the data analysis, discussion, and conclusions and providing copies of instruments in the appendix (A-C) (Figure 8.1).

8.3   Epistemological Approach towards the Delphi Study

The Delphi study has been identified to be pragmatic in nature. It can either be used as qualitative or quantitative research. It can be referred to as a flexible research method that can be applied in qualitative, quantitative or mixed research methods (Skulmoski et al., 2007). Due to its flexibility, the Delphi method possesses and can

Research problem:	Clearly defined
Research rationale:	Topic and method justification
Literature review:	Topic understudy
Methodology:	Data collection: clear explanation of the Delphi method employed
	Rounds: number employed, outline of each
	Sample: experts selection process and characteristics described in detail
	Reliability and validity issues identified
	Statistical interpretation: guidelines for the reader
	Ethical responsibilities: towards 'expert' sample and research community
Data Analysis:	Response rate for each round
	Round 1: presentation of total number of issues generated
	Round 2: presentation of results indicating the strength of support
	Further rounds (if applicable): presentation of results
Discussion :	Issue of consensus
and conclusions	Interpretations of consensus gained/not gained
	Direction of further research leading from conclusions
Appendices:	Copy of each round questionnaire illustrated.

Figure 8.1 Reporting a Delphi study (Hasson et al., 2000)

be applied both in the interpretivist and positivist research paradigms; hence, it is versatile. According to Skulmoski et al. (2007), other features inherent in Delphi that make it pragmatism aligned are its inexpensive nature and can be disseminated traditionally or electronically. It seeks a purposive sample with specific expertise on a subject, and unlike other research designs, it lacks complexity. Furthermore, it can be regarded to be rooted in the pragmatic paradigm because (Linstone & Turoff, 2002) describes it as a last resort for resolving complex problems where there exist inadequate models for the purpose. Hence it provides a best-fit approach to understanding the problem.

Taking this a step further to adequately understand the philosophical basis for the Delphi method, the works of Linstone and Turoff (2002) and Helmer and Rescher (1960) were examined. These studies classified the Delphi method to be Lockean and Leibnizian in nature. Lockean philosophy believes truth is experiential, thus empirically proven. However, this empirical content is tested through human observations. Also, it does not believe in theoretical considerations but direct observations. Meanwhile, the Leibnizian philosophy believes in the theoretical considerations to inquiries, and hence it is deductive. Through the Lockean philosophy, Delphi provides rich data to solve complex problems.

From the aforementioned, the Delphi method was adopted to achieve data from selected experts in construction digitisation. This ensured the gathering of data through expert consensus based on their contribution. Furthermore, the Delphi method was chosen for its pragmatic philosophical perspective because the problem of BIM maturity in the SACI is complex and without an existing framework in the

SACI. Hence the adoption of Delphi to achieve the data gathering and achieve robust data on the research study. This study adopted the electronic dissemination of the Delphi instruments; this aligns with the literature.

8.4   Designing and Executing the Delphi Study

Since the use of Delphi by the RAND Corporation and the publication about the method (Helmer & Rescher, 1960), there has been a replete of studies adopting the methodology. Although the method is considered void of complexities, it must be systematically executed, unlike other research designs. The design and execution require a systematic approach in order to achieve the right results.

To adequately understand the design of the Delphi process, the report from the original "project Delphi" at the RAND corporation was examined. The project Delphi report by Dalkey and Helmer (1963) outline the Delphi method involving the repeated expert questioning of experts without physical confrontation to include the administration of questionnaires and a follow-up interview to confirm responses to a seven-man expert panel. A survey of the literature suggests that the Delphi method ever since can be said to contain the same methodology. Milevska-kostova and Dunn (2014) stated that conventional Delphi consists of ten steps:

1 Formation of a Delphi team to undertake and monitor the project (researcher or the research team).
2 The selection of one or more panels to participate in the exercise. Typically, these are experts in the investigation area.
3 Development of the first round Delphi questionnaire.
4 Testing the questionnaire for proper wording (e.g., ambiguities and vagueness).
5 Transmission of the first questionnaires to the panellists.
6 Analysis of the first-round responses.
7 Preparation of the second-round questionnaires (and possible testing).
8 Transmission of the second-round questionnaires to the panellists.
9 Analysis of the second-round responses. (Steps 7 to 9 are reiterated as long as desired or necessary to achieve stability in the results.)
10 Preparation of a report by the analysis team to present the conclusions of the exercise.

These ten steps can easily be reworked and compressed into the study problem's definition, panel selection and panel size determination, and conducting Delphi rounds. The conduct of the Delphi process includes the administration of the questionnaire, which does not have a predefined number of rounds, the analysis of results and feedback. The rounds are stopped when the result reflects a consensus among the experts. The outlined design is seen in various studies (Aigbavboa, 2013; Gossler et al., 2019; Somiah, 2019). This study aligned with the Delphi method design adopted in these studies.

The various steps used in this study are discussed below.

PHASE 1 – DELPHI QUESTION DEVELOPMENT

This stems from the identification of the problem and the reason for conducting the research. Developing questions that achieve the aim of the research is central to the entire Delphi process. Achieving this phase also assist in the selection of the right experts for the study. Table 8.1 was adopted for this study to achieve the correct wording and ask the right questions.

PHASE 2 – CHOICE AND SELECTION OF DELPHI EXPERT PANEL

This phase is very central to the success of the Delphi method after the articulation of the study problem. The selection and choice of panel experts are important as their aggregated input determines the study's outcome. As earlier stated, experts have deep knowledge and experience in a particular field of study. In choosing experts, the determination of the level of expertise is important. The determination of this and the choice of experts must not be made based on acquaintance. Although Welty (1973) is still controversial in literature, this has been resolved by introducing expertise qualifications that determine panel membership. The qualifications

Table 8.1 Framing the Delphi questions

Main Delphi questions	Wording for the study
What is the motivation for the study?	The low maturity of BIM diffusion in SACI coupled with the struggle of many stakeholders with the successful BIM adoption motivated this study. There exist no systematic framework for BIM adoption in SACI. Therefore, this study is necessary to develop a BIM maturity framework for the SACI.
What is yet to be known that requires finding out by the study?	The SACI is replete with research efforts focused on the barriers to BIM diffusion and benefits. However, none has been able to develop a model to walk stakeholders through BIM adoption and implementation. This study is poised to deliver a maturity model enabling a systematic diffusion of BIM in SACI.
How do the Delphi results impact BIM maturity in SACI?	The Delphi results produce an in-depth and robust result that enables conceptualising BIM maturity in SACI.

are determined by the respective disciplines of the prospective panel members and not the researcher (Avella, 2016). This is because qualification and membership must be objectively based on qualifications and not subject to the researchers' feelings and preferences.

Another important criterion that must be ensured in making selection choices is impartiality. This is important as bias by panel members will provide a biased result. Ordinarily, the result is an aggregate of subjective opinions; hence impartiality by panel members must be discouraged through the choice (Hasson et al., 2000). Therefore, there must be a list of criteria for all panel members, including scholarly publications and professional bodies (Avella, 2016).

Furthermore, homogeneity is considered to stifle creativity and robustness in Delphi. A heterogeneous panel of Delphi experts is considered creative and good for the final results of a Delphi (Okoli & Pawlowski, 2004). A heterogeneous panel also ensures diverse perspectives and checks validity (Linstone & Turoff, 2002; Turoff, 1970). A heterogeneous panel is made up of individuals possessing expertise in a particular subject area but drawn from different stratifications professionally and socially (Clayton, 1997).

Consequently, based on the panel selection criteria and guidelines in literature, this study adopted a list of qualifications for experts for this study. Experts are expected to meet at least five of the outlined criteria:

1 **Practice residency**: Resides and practices in the South African construction industry for at least five years.
2 **Academic/Professional Qualification**: Has an earned degree (National Diploma/B-Degree/M-Degree/PhD) related to any aspect of construction practice. Postdoctoral training, certification employment/experience focusing on BIM diffusion.
3 **Experience**: Has a history of/currently is performing consultation/contracting services for a South African organ of State, individuals, businesses, agencies, companies, and/or organisations related to low income or other sustainable development or human settlement context.
4 **Employment:** Currently serves (or has previously served) in a professional or voluntary capacity (e.g., at the place of employment – institution, business, agency, department, company) as supervisor or manager of an establishment involved in construction and BIM diffusion in South Africa.
5 **Influence and Recognition:** One who has served/currently is serving as a peer-reviewer for one or more manuscripts received from a journal editor before its publication in the primary literature, focusing on BIM diffusion and construction digitisation manuscript(s).
6 **Authorship**: Is an author/co-author of peer-reviewed publications in construction digitisation emphasising South Africa; has prepared and presented papers at conferences, workshops, or professional meetings focusing on the construction industry and BIM diffusion.

7 **Research:** Has submitted one or more proposals to or has received research funding (grant/contract) from national, provincial, local government, regional, and/or private sources that support research efforts on BIM diffusion in the South African construction industry.

8 **BIM enthusiast**: Is passionate about the diffusion of BIM in the South African construction industry. He must belong to an organisation that has organised, prepared, and successfully presented one or more BIM training, seminars, or workshops. The workshop or course must have been targeted at the construction industry stakeholders.

9 **Membership**: Be a member of a construction industry professional body. Should be the representative of a professional body so that their opinions may be adaptable or transferable to the population.

10 **Willingness:** Panel members must be willing to participate in the entire Delphi study fully

The adoption of five selection criteria is in tandem with the study of Aigbavboa (2013) instead of the stringent four used by Somiah (2019). However, the study considered that five criteria are very important for all experts. The first of these is willingness. The Delphi process is a rigorous one and requires the participation of experts irrespective of the number of iterations. The five criteria are academic qualification, working experience, professional body membership, BIM/digitisation knowledge and occupational environment. These criteria are set to have currency in expert knowledge hence valuable contribution. Also, it follows the study by Schlecht et al. (2021) on blockchain where experts know-how in the field, work experience, interest in technology, occupation and their function in the organisation were determinants for expert selection.

Experts for this study were recruited via emails after sourcing for them via a network, LinkedIn, and BIM/digitisation research publication spaces. A brief overview of the study was sent to a total of 24 prospective experts, and their response was solicited with a brief qualification questionnaire. Upon the receipt of willingness feedback, a total of 14 experts met the qualification criteria. The Qualified experts were subsequently sent the round one Delphi instrument. This instrument contained a detailed study description and instructions. Also, it contained a structured questionnaire with closed and open-ended questions. The questionnaire design for round one was based on an extensive literature review about BIM diffusion, implementation, and maturity. It is worthy of note that one expert dropped out and could not keep up with the two rounds of the study.

There have been several debates on the adequacy of panel size for a Delphi study. It can be seen that there is no standard rule but ranges. Below are different studies with different ranges and adoptions for the panel size

From Table 8.2, it is evident that the panel size for Delphi depends on the type of study and the population type if homogenous or heterogeneous. Also, studies sometimes do not end with the same number of experts at the end of all

Table 8.2 Sample size in literature

Study	Panel size
Aigbavboa (2015); Kirschbaum et al. (2019); Aigbavboa (2013); Somiah (2019)	15
Hallowell and Gambatese (2010)	8–12*
Kattirtzi and Winskel (2020)	127 first round, 69 second round
(Holey et al., 2007b)	12
Avella (2016)	10–100*
Gossler et al. (2019)	31
Chan et al. (2015)	105 first round, 93 second round
De Loë et al. (2016)	Less than 10–1000*
Kluge et al. (2020)	43 first round, 38 second round
Jeon et al. (2015)	28 first round, 25 second round, 23 third round
Schlecht et al. (2021)	74 first round, 51 second round
Clayton (1997)	5–30*
Tengan (2018)	11

*Review papers

iterations. This also brings to the fore the pragmatic nature of the Delphi method being reflective in the sample size determination. For this study, as earlier stated, a total of 13 experts. The panel is considered heterogeneous as it consists of different professionals drawn from different social/professional stratifications. This is considered adequate given the panel population (Clayton, 1997) and the sizes in Table 8.2. The Delphi experts for this study consists of nine industry professionals: four architects, one principal partner of an architectural firm/architect, two quantity surveyors, and two architects who are BIM lead and a head of the BIM department in architectural firms. Also, four academics consisting of two Professors and two lecturers. It thus provides a heterogeneous. They all reside and practice in South Africa and meet the academic qualification. The least is a Bachelor's degree, and the highest is a PhD. All experts are in active service, and they all possess the SACI and BIM knowledge. All experts for the study met at least five of the qualification criteria.

PHASE 3 – CONDUCTING THE DELPHI ITERATIONS

8.5   Data Collection through Delphi

One of the features of the Delphi method is the iterative process and the controlled feedback nature. The number of rounds is not predefined in the literature; the study is stopped when the required rounds for the study are reached, or consensus is achieved (Hallowell & Gambatese, 2010). First round normally introduces the research study to the experts and offers them the opportunity to give their opinion and propose new opinions. The subsequent rounds seek an agreement on the factors in the study and increase concurrent validity (Hasson et al., 2000). It also allows for modification of their judgements and presents them with a larger information base than the first round (Linstone & Turoff, 2002). Iterative rounds can be between one and three rounds (Hallowell & Gambatese, 2010), two rounds (Gossler et al., 2019; Kirschbaum et al., 2019; Schmalz et al., 2021), and three rounds (Mokkink et al., 2020; Schlecht et al., 2021), among others, until stability of results is achieved (Milevska-kostova & Dunn, 2014).

It should be noted that experts sometimes drop off between rounds due to the demanding nature of the Delphi process. Hence the number of rounds is pragmatic (Chalmers & Armour, 2005). An increased number of rounds leads to reduced performance (Linstone & Turoff, 2002); hence the goal is not to perform many rounds of the Delphi process but to achieve a stable result among the panel.

The Delphi rounds for this study adopted the two rounds iterative process. The rounds were stopped when consensus was observed among the panel members on the BIM maturity factors in SACI. A structured Delphi questionnaire (Appendix B) was sent out to panel members via email; follow-up and subsequent round was conducted electronically. Adopting electronic Delphi allowed for prompt and faster reach of participants and hence saved time in communication. The first-round Delphi questionnaire was designed based on an extensive literature review on BIM diffusion and implementation. Experts were requested to attend to the questions based on their expertise and experience.

After the first round, the subsequent round questionnaire was shaped based on the responses from the experts in the first round. The first-round questionnaire contained factors on BIM maturity pulled from extant literature, and the questions were both closed-ended and open-ended. This allowed the panel members to suggest factors that, in their opinion, have not been captured. The second-round questionnaire allowed panel members to review their previous round responses and adjust it where necessary based on the group median. It is worth knowing that the second-round questionnaire contained closed-ended questions. For both rounds, the responses were analysed using the measures of central tendency to measure the degree of consensus among the panel members on the BIM maturity factors in SACI.

The questionnaire consists of a rating scale and instructions requesting panel members to rate the likelihood of individual attributes and the impacts of the sub-attributes in predicting BIM maturity in SACI. Table 8.3 shows the influence rating scale in the Delphi questionnaire.

During the controlled feedback process between rounds, the result from the rounds was analysed statistically using the median score. The median is one of the measures for determining the consensus among the panels (Holey et al., 2007). Other measures are K value, percentage, and mean, among others. The median was adopted to show consensus because it considers every response in achieving its results. The median for each attribute was calculated, and the result was sent back

Table 8.3 Influence scale (probability in percentage)

0–10%	11–20%	21–30%	31–40%	41–50%	51–60%	61–70%	71–80%	81–90%	91–100%
1	2	3	4	5	6	7	8	9	10

to experts to allow them to review their responses in light of the group median. Experts can either maintain their responses in round one, modify it or go with the group median. Experts whose new responses were one unit (10%) above or below the group median were requested to provide a reason for the outlying response in the field provided.

When the results of the first round were analysed and returned to the experts for their input in the second round, the results from the second round were also analysed, and it was observed that consensus was reached. Hence the study adopted two rounds of Delphi; therefore, the third round was not required. It is worthy of note that throughout the Delphi process, all the principles of the Delphi methodology were upheld.

8.6 Specific Objectives of the Delphi

Literature is filled with BIM diffusion, implementation, and adoption factors in the SACI. However, a systematic framework for BIM adoption in the SACI is non-existent in the literature. Thus, a reliable measure that will provide a robust understanding of BIM diffusion and maturity is needed. Delphi method provides this as a research design. Therefore, it was employed in gathering robust data in understanding BIM maturity in the SACI. To this end, different questions were developed to be answered using the Delphi research method. The objectives to be achieved by the Delphi survey for this study are:

DS01 To identify the attributes (main and sub-attributes) that determine BIM maturity in the SACI.

DSO2 To determine the BIM maturity level of the SACI

DSO3 To determine the critical success factors of BIM diffusion in the SACI

DSO4 To identify peculiar BIM maturity attributes to the SACI

The objectives outlined above is backed by the philosophy of achieving coherence in the BIM maturity in the SACI. Also, unearthing the hidden attributes that were hitherto neglected in BIM diffusion research in SACI. Consequently, the objectives achieved the following outcomes:

– Determining the main factors and attributes that critically influence the determinants of BIM maturity in SACI.
– Developing a holistic conceptual BIM maturity model for SACI.

8.7　Computation of Data from Delphi Survey

This study adopted the use of Microsoft Excel software in the analysis of Delphi round results. To determine the level of agreement within the panel, the study used the median in the analysis of the first-round survey questionnaire. Also, the interquartile deviation (IQD) was also adopted. This analysis is based on extant literature (Holey et al., 2007a) and applied in various studies (Aigbavboa, 2013; Somiah, 2019; Tengan, 2018). These studies also apply the absolute deviation of the group medians. This is the deviation between response in a data set and a given point (Aigbavboa, 2013; Hallowell & Gambatese, 2010). For this study, the point of reference is the median, the formula for calculating the absolute deviation is given below:

$$D_i = (x_i - m(X))$$ *Equation 1*

Where:
D_i = Absolute deviation
X_i = Panellist rating
$m(X)$ = Measure of central tendency

For each attribute and sub-attribute, the influence on BIM maturity was calculated based on expert opinions. This analysis is presented as percentages and numbers in tables and pictorially using bar charts.

8.8　Obtaining Consensus from the Delphi Process

This has been described as one of the most difficult stages in the Delphi process (Hallowell & Gambatese, 2010). Different studies have adopted different methods to achieve consensus in Delphi studies. This might not be unconnected with the different format rating scales and the consensus thresholds adopted in these studies; these factors have been observed to affect consensus (Lange et al., 2020). Expectedly, various studies have adopted different yardsticks to determine consensus; this includes the median, standard deviation, absolute deviation, per cent agreement, IQR, and Kendall's W coefficient, among others. Although, most of these measures have been observed to be inconsistently used in different studies. For example, Kirschbaum et al. (2019) observed this in IQR/IQD adoption in the determination of consensus by Delphi studies. Their study adopted a combination of IQR≤1 and a median cut-off of ≥4.

Similarly, Gossler et al. (2019) adopted a combination of indicators, are IQR and r_{wg}. Studies have also adopted only one indicator in determining consensus; for instance, Okoli and Pawlowski (2004) and Chan and Chan (2012) adopted Kendall's W in determining consensus. Kluge et al. (2020) adopted the use of IQR to determine consensus. It can therefore be inferred that there are no standard rules for the determination of consensus in Delphi studies. However, according to Holey et al. (2007), for consensus to be achieved by aggregation of panel judgements, there has to be a move of opinions to a subjective level of central tendency or the consistency or stability of responses between rounds.

For this study, achieving consensus is considered appropriate to end the Delphi rounds. To this end, the study adopted the calculation of the group median and IQD for each attribute to determine consensus. The consensus was determined through the use of IQD not more than one unit. It is worthy of note that the study adopted a 10 Likert scale to measure the influence of each attribute on the BIM maturity of SACI. The absolute deviation was calculated using Equation 1, as earlier stated. The IQD is the difference between the 75th and 25th percentiles. It can be expressed as IQD=Q3-Q1. The IQD gives an absolute value which is considered robust statistics (Aigbavboa, 2013). Smaller figures indicate a higher degree of consensus, and the reverse is the case for higher figures.

The study, therefore, determined consensus using the following interpretation and guides for the statistical analysis.

1 If more than 60% of responses are either positive or negative for certain attributes.
2 The average absolute deviation (calculated from equation 1) was not more than one unit.
3 Where the IQD was less than 1.00. This translates that the items with IQD = 0.00 were considered to reflect high consensus.

Consequently, the scales of consensus adapted for this study are:

1 Strong consensus – median 9-10, mean 8-10, interquartile deviation (IQD) ≤1 and ≥80% (8-10);
2 Good consensus – median 7-8.99, mean 6-7.99, IQD≥1.1≤2 and ≥60%≤79% (6-7.99); and
3 Weak consensus – median ≤ 6.99, mean ≤5.99 and IQD≥2.1≤3 and ≤ 59% (5.99).

8.9   Reliability and Validity of the Delphi Method

The concepts of reliability and validity in research are very important as they are central to accepting the results. Reliability represents the extent to which a particular research instrument or procedure will produce the same result under the same conditions at different times. However, this is hard to achieve in Delphi studies due to the subjective nature of the responses (Aigbavboa, 2013). One of the proposed ways of preserving validity proposed in the literature is the heterogeneity of the panel (Linstone & Turoff, 2002). In addition, the Delphi method must comply with four criteria to ensure its credibility; these are credibility (truthfulness), fittingness (applicability), audibility (consistency) and confirmability (Hasson et al., 2000). Validity is ensured through the process of controlled feedback where assumptions are challenged (Hasson et al., 2000).

For this study, reliability and validity were ensured firstly by adhering to all the principles of the Delphi method. This includes maintaining anonymity among panellists and ensuring a controlled feedback system, among others. Through the iterative rounds of the Delphi process, reliability was ensured. The expert panel's ability to achieve consensus over the rounds is thus interpreted to achieve reliability for this study.

Secondly, the internal and external validity of the Delphi survey for this study was determined through the anonymity of the panel members. This ensured that the "bandwagon effect' was eliminated from the study. Furthermore, the individual method of communication with experts and the availability of opportunities to review opinions freely by experts also increased the internal validity of the Delphi method for this study. Secondly, the careful selection of experts to achieve heterogeneity also ensured the external validity of the process. Consequently, the result can be generalised to a larger population.

8.10   Results of the Delphi Process

8.10.1 Background to the Delphi Survey

As previously stated, the objectives of the Delphi survey for this study are as follows:

DS01   To identify the attributes (main and sub-attributes) that
determine BIM maturity in the SACI.
DSO2   To determine the BIM maturity level of the SACI
DSO3   To determine the critical success factors of BIM diffusion in the SACI
DSO4   To identify peculiar BIM maturity attributes to the SACI

These objectives were taken from the overall study objectives. The Delphi survey aimed to achieve a consensus from an aggregate of expert opinions on the BIM diffusion and implementation in the SACI. Therefore, the Delphi survey objectives identified the key influencing factors regarding BIM maturity in the SACI. Consequently, it contributed to the development of the holistic and conceptual BIM maturity model for the SACI.

The Delphi survey for this study involved a 13 experts panel drawn from different strata in academia and the construction industry. The five criteria are academic qualification, working experience, professional body membership, BIM/digitisation knowledge, and occupational environment. These criteria are set to have currency in expert knowledge hence valuable contribution. Also, it follows the study by Schlecht et al. (2021) on blockchain where experts' know-how in the field, work experience, interest in technology, occupation, and their function in the organisation were determinants for expert selection.

The Delphi experts for this study consists of nine industry professionals (Four Architects, one principal partner of an Architectural firm/Architect, two Quantity surveyors, two Architects who are a BIM lead and a head of the BIM department in Architectural firm). Furthermore, the Delphi experts for this study comprise four academics comprising two Professors and two lecturers. It thus provides a heterogeneous. They all reside and practice in South Africa and meet the academic qualification. The least is a Bachelor's degree, and the highest is a Ph.D. All experts are in active service, and they possess the SACI and BIM knowledge. All experts for the study met at least five of the qualification criteria.

An extensive review informed the first-round questionnaire of extant literature. However, the content of the second-round questionnaire is based on the responses of the experts from the first round. The questions in the first round consisted of closed and open-ended questions, expert responses to these questions were analysed and used for round two. It is worthy of mention that a vital function of the first round is that it allowed the experts to brainstorm and add their voices to the study.

The measure of central tendency was applied to analyse the responses for the first round. This allowed for measuring the level of consensus reached among the experts on the BIM maturity factors in the SACI. These were communicated with the experts for the second round with new factors proposed by the experts. The second round allowed experts to review their responses. Also, it allowed experts to rate the influence and respond to the proposed factors by the experts in the first round. The analysis of the second-round results showed that consensus had been reached among the experts. The second-round Delphi results were analysed using the median, mean, standard deviation, and the IQD. Therefore, there was no need for a third round of the Delphi survey. The second-round results were employed in developing the list of factors influencing the BIM maturity model in the SACI and in the extension of the conceptual framework.

8.11 Delphi Study Findings

8.11.1 Delphi Specific Objectives

To adequately calculate consensus among expert opinions and interpret it, different analyses were conducted as earlier mentioned. Experts' response in each round depends on an influence rating scale of 1–10 (1 = low influence and 10 = high influence). Experts are expected to rate each attribute based on the 10-ordinal scale provided. Hence, the consensus is based on the levels of influence as depicted by the expert response. For the second round, the result from the first round was presented to the experts using the median to show the group response. The median takes into consideration all responses on the scale in its calculation. Thus, it presents an objective position of the group response. The second round allowed experts to review their responses against the group median; thus, they can either maintain their original responses, accept the group median, or indicate a new response on the questionnaire. The questionnaires for the two rounds are presented in the appendix.

The following categorisation was adopted in this study to interpret the Delphi result analysis adequately:

i Strong consensus: median 9-10, mean 8-10, IQD≤1 and ≥ 80% (8-10)
ii Good consensus: median 7-8.99, mean 6-7.99, IQD≥1.1≤2 and ≥ 60%≤79% (6-7.99)
iii Weak consensus: median ≤6.99, mean ≤5.99 and IQD≥2.1≤3 and ≤59% (5.99)

The influence scale was categorised as follows:

i VHI: 9 – 10
ii HI: 7 – 8.99
iii MI: 3 – 6.99

It should be noted that this study only chose factors with strong consensus using the IQD of 1 as the basis for this decision.

DS01, DSO3, and DSO4 – To identify the attributes (main and sub-attributes) that determine BIM maturity in the SACI, to determine the critical success factors of BIM diffusion in the SACI, and to identify peculiar BIM maturity attributes to the SACI, respectively.

Experts were asked to rate the influence of the following attributes on BIM maturity in the SACI.

8.11.2 *People Attributes*

This main attribute measured the human actors involved with implementing the processes and the use of BIM technology in work environments (organisation and project) to achieve BIM maturity in the South African construction industry. The sub-attributes were divided and measured at the micro, meso, and macro levels.

The experts were presented with 38 people attributes in the first round to rate their influence on the BIM maturity in the SACI. Three attributes, unstable economic development, high rate of corruption, and "every man for himself" syndrome," were added by experts at the end of the first round. It was observed that five constructs were considered not to influence the BIM maturity of the SACI. These factors are as follows:

– Actor role definition with a median of 7, SD of 2, mean of 6.31, and IQD of 2
– Actor role qualification with a median of 7, SD of 2, mean of 6.31, and IQD of 2
– The technological mindset with a median of 9, SD of 1, mean of 8.23, and IQD of 2
– Firm financial resources with a median of 8, SD of 1, mean of 7.85, and IQD of 2

A critical look at these four constructs reveals that technological mindset highly influences BIM maturity, but experts did not achieve a consensus on this attribute. Also, there was no consensus on the influence of BIM actors on the BIM maturity of the SACI. However, there were three new attributes proposed by experts in the first round. These constructs are:

– Unstable economic development
– High rate of corruption
– Everyman for himself syndrome

Out of these three proposed attributes, high rate of corruption did not achieve consensus. The addition of the three proposed attributes brought the people attribute for this study to forty-one. Generally, people attributes were rated to have a high influence on the BIM maturity of the SACI. The only point of divergence is reaching a consensus.

	Code	*PEOPLE ATTRIBUTE (PEA)*	*Median*	*SD*	*Mean*	*IQD*
Micro-level	PEA1	Competency sets	7	2	6.85	1
	PEA2	Actor role definition	7	2	6.31	2
	PEA3	Actor role qualification	7	2	6.31	2
	PEA4	Resistance to change	9	1	8.77	0
	PEA5	Change readiness	8	1	8.15	0
	PEA6	Willingness to share information	8	1	8.38	1
	PEA7	Unequal adoption readiness	7	2	7.08	1
	PEA8	Individual awareness level	9	2	8	1
	PEA9	Other stakeholders BIM involvement	8	1	7.92	0
	PEA10	Client support and involvement	9	1	8.62	1
	PEA11	Technological mindset	9	1	8.23	2
	PEA12	Opinion leaders influence	8	1	7.38	1
Macro-level	PEA13	Multiple stakeholders with diverse interests/need	7	1	7.08	1
	PEA14	Conservative clients	8	1	7.69	0
	PEA15	Government mandate	9	2	8	1
	PEA16	Government financial subsidy	6	2	6.08	1

(Continued)

	Code	PEOPLE ATTRIBUTE (PEA)	Median	SD	Mean	IQD
	PEA17	BIM training	9	2	8.85	1
	PEA18	BIM skills	9	1	9	0
	PEA19	BIM Knowledge	9	1	8.92	0
	PEA20	Leadership style	8	1	8	0
	PEA21	Organisational culture	8	1	8.15	0
	PEA22	Management framework	7	1	7.23	0
	PEA23	Research and development efforts	8	1	7.69	0
	PEA24	Strategic alignment	8	1	7.85	0
	PEA25	Reward system	7.5	1	6.92	1.5
	PEA26	Actual impacts of BIM	8	2	7.62	1
Meso level	PEA27	Financial resources	8	1	7.85	2
	PEA28	Management support	9	2	8.15	0
	PEA29	Employee incentive	7	1	6.54	1
	PEA30	Unstable workforce	6	1	6.62	1
	PEA31	Vision and goals	8	2	7.92	1
	PEA32	Degree of compliance with organisational objectives	8	1	7.69	0
	PEA33	Information management within the organisation	7.5	1	7.54	0.5
	PEA34	Aligning and managing resources and teams to focus on digital services	8	1	7.77	0
	PEA35	Mission and objectives	7	2	6.92	0
	PEA36	Awareness level	7	1	6.92	0
	PEA37	Recruitment challenge	6	1	6.15	0
	PEA38	Organisational structure	7	1	7.25	1
New constructs	PEA39	Unstable economic development	7	1	7	1
	PEA40	High rate of corruption	7	3	6.86	1.5
	PEA41	"Every man for himself" syndrome	8	3	7	0.5

8.11.3 *Technology Attributes*

This study's technological attributes include the hardware, software requirements, and network integration for BIM usage and maturity in an environment (organisation and project). This measures the infrastructure and other attendant factors required for BIM maturity. This main attribute is measured under the following categories: software, physical infrastructure, and data management.

A total of twenty-six attributes were presented to experts to rate the level of influence on the BIM maturity of SACI. At the end of the first round, the experts added three attributes, thus bringing the total number of attributes to twenty-nine. Nine out of the original attributes were observed not to achieve consensus, and two of the proposed new constructs did not achieve consensus.

The following sub-attributes did not achieve consensus under the technology attribute:

- BIM functions with a median of 6, SD of 2, mean of 7.08, and IQD of 2
- Availability of software for selection with a median of 7, SD of 1, mean of 7.77, and IQD of 2

- BIM applications with a median of 7, SD of 2, mean of 7.54, and IQD of 2
- Graphics with a median of 7, SD of 2, mean of 7.23, and IQD of 2
- Cost of software renewal with a median of 8.5, SD of 1, mean of 8.31, and IQD of 2
- Affordability of cloud-based technology with a median of 8, SD of 1, mean of 8, and IQD of 2
- Infrastructure with a median of 7, SD of 2, mean of 6.62, and IQD of 2
- Data security with a median of 8, SD of 2, mean of 7.92, and IQD of 2
- Availability of database technology with a median of 8, SD of 2, mean of 8, and IQD of 2
- Cybersecurity with a median of 7, SD of 2, mean of 7.69, and IQD of 2
- Cost of technology infrastructure with a median of 8, SD of 1, mean of 7.86, and IQD of 1.5
- Innovations are developed country tailored with a median of 6, SD of 2, mean of 6.43, and IQD of 3

A critical look at these attributes shows that the cost of software renewal was rated to have a high influence but did not achieve consensus.

	Code	TECHNOLOGY ATTRIBUTES (TEC)	Median	SD	Mean	IQD
Software	TEC1	BIM functions	6	2	7.08	2
	TEC2	Availability of Software for selection	7	1	7.77	2
	TEC3	BIM applications	7	2	7.54	2
	TEC4	BIM tools interoperability	8	1	8.15	1
	TEC5	Complimentary technologies	7	1	7.23	0
	TEC6	Graphics	7	2	7.23	2
	TEC7	Cost of software renewal	8.5	1	8.31	2
	TEC8	Availability/affordability of cloud-based technology	8	1	8	2
	TEC9	Technical support by software vendors	8	1	8.15	1
	TEC10	Spatial capabilities	7	2	6.69	0
	TEC11	Geospatial capability	7	2	7.15	0
	TEC12	BIM networking establishments (e.g., intranets, extranets, and platforms)	7	2	7.54	1
Physical infrastructure	TEC13	Innovativeness	8	1	7.69	1
	TEC14	Hardware (equipment)	8	2	8.08	1
	TEC15	Hardware upgrade	8	1	8	0
	TEC16	Hardware availability	8	2	7.77	0
	TEC17	Infrastructure (e.g., relating physical space building)	7	2	6.62	2

(Continued)

	Code	TECHNOLOGY ATTRIBUTES (TEC)	Median	SD	Mean	IQD
Data management	TEC18	Data Security	8	2	7.92	2
	TEC19	Availability of database technology	8	2	8	2
	TEC20	Data exchange	8	1	8.38	1
	TEC21	Data storage	8	2	7.69	1
	TEC22	Data richness	8	1	7.92	1
	TEC23	Real-time data	9	1	8.62	0
	TEC24	Data access control	8	1	8.15	0
	TEC25	Intellectual property rights	8	1	8	0
	TEC26	Cybersecurity	7	2	7.69	2
New constructs	TEC27	Willingness by professionals to initiate BIM-applied projects without a mandate from the public sector	8	1	8.29	0.5
	TEC28	Cost of technology infrastructure	8	1	7.86	1.5
	TEC29	Innovations are developed country tailored	6	2	6.43	3

8.11.4 Process Attributes

The process attribute measures every activity carried out by teams during the life cycle of any building asset to achieve effective information management according to ISO 19650-2:2018. The original sub-attributes for this attribute were thirty-five in total. Experts proposed three new attributes for this main attribute. Under this construct, eleven did not meet consensus from the original attribute and two from the new attributes proposed by the experts. Under this attribute, the following sub-attributes did not meet consensus:

- Knowledge sharing process with a median of 8, SD of 2, mean of 8.23, and IQD of 2
- Handover processes between project bases with a median of 8, SD of 2, mean of 7, and IQD of 2
- Information documentation with a median of 8, SD of 1, mean of 8.23, and IQD of 2
- Delivery processes of BIM-related services with a median of 8, SD of 1, mean of 7.86, and IQD of 1.5
- Stakeholders management with a median of 8, SD of 2, mean of 7.08, and IQD of 3
- Data accuracy with a median of 9, SD of 2, mean of 8.23, and IQD of 2
- Level of detail with a median of 9, SD of 2, mean of 8.23, and IQD of 2
- Execution plan with a median of 8, SD of 2, mean of 7.31, and IQD of 3
- Change order management with a median of 8, SD of 2, mean of 7.46, and IQD of 2

- Information value assessment with a median of 8, SD of 2, mean of 7.46, and IQD of 2
- Specification with a median of 8, SD of 2, mean of 7.77, and IQD of 3
- Client refusal to pay for the BIM process-related services with a median of 8, SD of 2, mean of 7.83, and IQD of 2.3
- Client inability to pay for BIM process-related services with a median of 8, SD of 1, mean of 8, and IQD of 1.5

It is evident that two of them were rated to have a high influence on the SACI's BIM maturity; however, they did not achieve consensus. The sub-attributes, data accuracy, and level of detail were considered highly influential to the BIM maturity of the SACI.

Code	PROCESS ATTRIBUTES (PSA)	Median	SD	Mean	IQD
PSA1	Co-ordination processes between project phases	8	1	8.23	1
PSA2	Interaction coordination among multiple disciplines/stakeholders	9	2	8.46	1
PSA3	Information generation (e.g., quantity take-offs and weekly schedules)	8	1	7.85	0
PSA4	Delivery processes of BIM-related products	8	1	8	0
PSA5	Knowledge sharing processes	8	2	8.23	2
PSA6	Reuse procedures of BIM-related information and data	8	1	8.23	1
PSA7	Documentations of actually gained benefits on working processes through applying BIM	8	2	7.85	1
PSA8	Handover processes between project phases	8	2	7	2
PSA9	Information documentation	8	1	8.23	2
PSA10	Delivery processes of BIM-related services	8	1	7.92	2
PSA11	Documentations of actually gained impacts on working processes through applying BIM	7	1	6.85	0
PSA12	Collaboration among project team/stakeholders	8	1	8.15	1
PSA13	Information management	8	2	7.62	1
PSA14	Process standardisation	9	2	8.31	1
PSA15	Quality control	9	2	8.31	1
PSA17	Roles and responsibilities	8	2	7.77	0
PSA18	Communication among project team	9	1	8.69	0
PSA19	Stakeholders management	8	2	7.08	3
PSA20	Research and development	8	2	7.62	1
PSA21	Information sharing	9	1	8.62	0
PSA22	Data accuracy	9	2	8.23	2
PSA23	Level of detail	9	2	8.23	2
PSA24	BIM workflow	8	2	7.69	0
PSA25	Execution plan	8	2	7.31	3
PSA26	Appointment protocols	8	2	7.54	1
PSA27	Information delivery cycle	8	2	7.46	0
PSA28	Information requirements	8	1	7.54	0
PSA29	Information protocol	8	1	7.62	0

(Continued)

Code	PROCESS ATTRIBUTES (PSA)	Median	SD	Mean	IQD
PSA30	Mobilisation protocols	8	1	7.46	1
PSA26	Change order management	8	2	7.46	2
PSA27	Information value assessment	8	2	7.46	2
PSA28	Risk management	8	2	7.46	1
PSA29	Specification	8	2	7.77	3
PSA30	Project complexity	8	2	7.67	0.75
PSA31	Early involvement of the project team	10	2	9.15	1
PSA32	Contractual obligations involved with regard to ownership of the models	8	1	7.83	0
PSA33	Client refusal to pay for the BIM process-related services	8	2	7.83	2.25
PSA34	Client inability to pay for the BIM process-related services	8	1	8	1.5

8.11.5 Stakeholder Orientation Attributes

The stakeholder orientation attribute measures the perception and knowledge of the construction industry stakeholders regarding BIM. The sub-attributes under this attribute are classified under perceived usefulness, perceived ease of use, relative advantage, compatibility, trialability, observability, and complexity. This attribute consists of twenty-six sub-attributes in all. The experts did not propose any new sub-attribute under this main attribute. However, not all of the sub-attributes achieved consensus. Four sub-attributes that did not achieve consensus are:

- The convenience of the BIM process with a median of 8, SD of 2, mean of 7.92, and IQD of 2
- Individual conviction about BIM benefits with a median of 8, SD of 2, mean of 6.92, and IQD of 3
- Enhanced job output with a median of 8, SD of 2, mean of 7.15, and IQD of 2
- Ease of incorporation into existing work process with a median of 8, SD of 2, mean of 7.31, and IQD of 2

Unlike previous characteristics observed in some of the sub-attributes that did not achieve consensus, all the sub-attributes under the stakeholder orientation attribute did not achieve consensus and did not get a high influence rating from experts.

	Code	STAKEHOLDER ORIENTATION ATTRIBUTES (SOA)	Median	SD	Mean	IQD
	SOA1	Improvement of job satisfaction	7	2	6.31	1
Perceived usefulness	SOA2	Improvement of job outcomes	7	2	6.54	1
	SOA3	Improvement of job productivity	7	1	6.92	0
	SOA4	Efficiency	8	1	7.62	1

(Continued)

	Code	STAKEHOLDER ORIENTATION ATTRIBUTES (SOA)	Median	SD	Mean	IQD
Perceived ease of use	SOA5	Convenience of the BIM process	8	2	7.92	2
	SOA6	Ease of getting expected outcomes from BIM	8	2	7.85	1
	SOA7	Individual conviction about BIM benefits	8	2	6.92	3
	SOA8	Ease of adoption	7	2	7.15	1
Relative advantage	SOA9	Profitability	8	2	7.85	0
	SOA10	Enhanced corporate image	8	1	7.92	1
	SOA11	Satisfaction	8	1	7.31	1
	SOA12	Risk reduction	8	2	7.69	1
	SOA13	Enhanced job output	8	2	7.15	2
Compatability	SOA14	Compatability with values	7	2	6.92	0
	SOA15	Compatibility with existing beliefs	7	2	6.85	0
	SOA16	Compatibility with past experiences	7.5	2	7.33	0.25
	SOA17	Ease of incorporation into the existing work process	8	2	7.31	2
Trialability and observability	SOA18	Possibility of testing BIM in stages for a limited time before deciding on adoption	8	2	7.46	1
	SOA19	Experimenting and witnessing the benefits before an adoption decision	7	2	7	0
	SOA20	Visible benefits of BIM from others in the network	7	1	7.54	1
	SOA21	Witnessing the benefits before an adoption decision	8	2	7.77	1
	SOA22	Ability to test new ideas to ascertain learned benefits	7	1	6.92	1
Complexity	SOA23	Complication of use	8	2	7.83	0
	SOA24	Difficulty level of understanding BIM	8	1	8.15	1
	SOA25	The complexity of the BIM process implementation	8	1	8.23	1
	SOA26	Complexity of integrating the BIM process with the prevalent workflow	9	1	8.54	0

8.11.6 Ecosystem Regulation Attributes

Ecosystem regulations attribute measured the sets of standards and regulations articulating the alignment of the ecosystem with the BIM requirements. It creates a balance between the regulation structure and the technology. Experts were asked to rate the influence of these sub-attributes on the BIM maturity of the SACI.

Ten original sub-attributes measure this main attribute; however, one sub-attribute was added by the experts at the end of round one. The added sub-attribute is the "plethora of regulations in South Africa which negatively impacts the field of innovative development." Under this attribute, the sub-attributes that did not achieve consensus are:

- Government mandate with a median of 8, SD of 3, mean of 6.75, and IQD of 3
- Legal framework with a median of 7, SD of 2, mean of 6.17, and IQD of 2
- Guidelines of BIM-related information needs and information breakdown structure with a median of 8, SD of 2, mean of 7.5, and IQD of 2
- Noteworthy publications with a median of 7, SD of 3, mean of 6.17, and IQD of 2.25
- Data exchange standards with a median of 8, SD of 2, mean of 7.08, and IQD of 2.25

Code	ECOSYSTEM REGUALTION ATTRIBUTES (ERA)	Median	SD	Mean	IQD
ERA1	Government mandate	8	3	6.75	3
ERA2	Regulation and industry standards	9	3	8.08	1
ERA3	Other stakeholder regulations and guidelines	7	2	7.5	1.25
ERA4	Delivery benchmarks	8	2	7.75	0.75
ERA5	BEP adoption	6.5	1	6.67	0.5
ERA6	Third-party management	5.5	2	5.75	0.5
ERA7	Legal framework	7	2	6.17	2
ERA8	Guidelines of BIM-related information needs and information/model breakdown structure	8	2	7.5	2
ERA9	Noteworthy publications (NP)	7	3	6.17	2.25
ERA10	Data exchange standards	8	2	7.08	2.25
ERA11	Plethora of regulations in South Africa which negatively impacts the field of innovative development	8	1	7.6	1

8.11.7 *Effective BIM Training Attributes*

Effective BIM training attributes measured the various requirements for successfully integrating BIM into tertiary institutions as the incubator for industry professionals. The effective BIM training attributes contain a total of twenty-seven sub-attributes (twenty-six were originally presented, and experts proposed one at the end of round one). Experts ranked these attributes based on the influence on the BIM maturity of the SACI. From the twenty-seven sub-attributes. The following sub-attributes did not achieve consensus:

- Stand-alone courses with a median of 7.5, SD of 2, mean of 7.58, and IQD of 4
- Infrastructure with a median of 8, SD of 2, mean of 7.83, and IQD of 1.25

- Hardware upgrade with a median of 8.5, SD of 2, mean of 7.96, and IQD of 1.13
- Hardware (physical workspace) with a median of 8, SD of 2, mean of 7.75, and IQD of 1.25
- Teaching of BIM in skills-oriented tertiary institutions and not in theoretical institutions with a median of 9, SD of 2, mean of 8.2, and IQD of 2

Hardware upgrade and teaching BIM in skills-oriented tertiary institutions and not in theoretical institutions were rated to have great influence. However, they did not achieve consensus based on expert opinions.

	Code	EFFECTIVE BIM TRAINING ATTRIBUTES (EBT)	Median	SD	Mean	IQD
Structure	EBT1	Academic framework	8.5	2	8.08	1
	EBT 2	Pedagogy standards and benchmarks	8	2	7.67	1.5
	EBT3	Cross-curriculum teaching module	8.5	1	8.67	1.5
	EBT4	Standalone course	7.5	2	7.58	4
	EBT5	Updating existing courses	6.25	2	6.42	0.75
	EBT6	Interactive modules	9	1	9	0.25
	EBT7	BIM-based capstone project experience	9	2	8.67	0.25
Pedagogical strategies	EBT8	Instruction and lecture	7	1	7.5	0.5
	EBT9	Extracurricular studies	6.5	2	6.46	0.88
	EBT10	Case study	7.5	2	7.17	0.25
	EBT11	Group discussion	7	2	6.5	0.25
	EBT12	Field trips	7.5	2	7.42	0.75
	EBT13	Interactive simulations	8	2	7.67	0
	EBT14	Problem based	8	2	7.64	0
Tutor engagement	EBT15	From industry (industrial consultant)	9.5	1	9	0.25
	EBT16	Faculty members	6.5	2	6.33	0.75
	EBT17	Guest lecturer for a topic	8	2	7.67	0
Assessment methods	EBT18	Conventional assessment	6	2	6	0.75
	EBT19	Assessment by invited jurors	7	2	6.83	0.25
	EBT20	Assessment by invited professionals	8	2	7.17	0.75
	EBT21	Verbal/written	5	2	5.33	0.25
Resource	EBT22	Infrastructure (e.g., internet and other facilities)	8	2	7.83	1.25
	EBT23	Hardware (e.g., equipment purchasing)	8.5	2	8.38	0.75
	EBT24	Hardware upgrade	8.5	2	7.96	1.13
	EBT25	Hardware (physical workspace)	8	2	7.75	1.25
	EBT26	Software	9	1	9.08	0

(Continued)

	Code	EFFECTIVE BIM TRAINING ATTRIBUTES (EBT)	Median	SD	Mean	IQD
New constructs	EBT27	Teaching of BIM in skills-oriented tertiary institutions and NOT in theoretical institutions	9	2	8.2	2

8.12   Summary

This chapter described the entire Delphi process and how it is executed. It provided insight into the process, panel constitution, and result analysis, among others. It also covered the validity and reliability of the Delphi process. This study conducted a two-round Delphi survey results based on experts' opinions. The sub-attributes that did not achieve consensus and those that did were highlighted and discussed.

References

Aigbavboa, C. (2015). A Delphi Technique Approach of Identifying and Validating Subsidised Low-Income Housing Satisfaction Indicators. In *International Conference of Korean Housing Association, OTMC Conference*, 567–574.

Aigbavboa, O. C. (2013). *An Integrated Beneficiary Centred Satisfaction Model For Publicly Funded Housing Schemes In South Africa*. University Of Johannesburg, Johannesburg, South Africa.

Avella, J. R. (2016). Delphi Panels: Research Design, Procedures, Advantages, and Challenges. In *International Journal of Doctoral Studies* (Vol. 11). http://www.informingscience.org/Publications/3561

Chalmers, J., & Armour, M. (2005). The Delphi Technique. *Education The Obstetrician & Gynaecologist*, 7, 120–125. https://doi.org/10.1007/978-981-10-5251-4_99

Chan, A. P. C., Lam, P. T. I., Wen, Y., Ameyaw, E. E., Wang, S., & Ke, Y. (2015). Cross-Sectional Analysis of Critical Risk Factors for PPP Water Projects in China. *Journal of Infrastructure Systems*, 21(1). https://doi.org/10.1061/(ASCE)IS.1943-555X.0000214

Chan, D. W. M., & Chan, J. H. L. (2012). Developing a Performance Measurement Index (PMI) for Target Cost Contracts in Construction: A Delphi Study. *Construction Law Journal (CLJ) (Final Accepted Manuscript)*, 28(8), 590–613.

Clayton, M. (1997). Delphi: A Technique to Harness Expert Opinion for Critical Decision-Making Tasks in Education Education of Students with High Support Needs View Project. *Educational Psychology*, 17(4). https://doi.org/10.1080/0144341970170401

Dalkey, N., & Helmer, O. (1963). An Experimental Application of the DELPHI Method to the Use of Experts. In *Management Science* (Vol. 9, Issue 3, pp. 458–467). https://doi.org/10.1287/mnsc.9.3.458

De Loë, R. C., Melnychuk, N., Murray, D., & Plummer, R. (2016). Advancing the State of Policy Delphi Practice: A Systematic Review Evaluating Methodological Evolution, Innovation, and Opportunities. *Technological Forecasting & Social Change*, 104, 78–88. https://doi.org/10.1016/j.techfore.2015.12.009

Devaney, L., & Henchion, M. (2018). Who is a Delphi 'Expert'? Reflections on a Bio-economy Expert Selection Procedure from Ireland. *Futures*, 99, 45–55. https://doi.org/10.1016/j.futures.2018.03.017

Eubank, B. H., Mohtadi, N. G., Lafave, M. R., Wiley, J. P., Bois, A. J., Boorman, R. S., & Sheps, D. M. (2016). Using the Modified Delphi Method to Establish Clinical Consensus for the Diagnosis and Treatment of Patients with Rotator Cuff Pathology. *BMC Medical Research Methodology*, *16*(1), 1–16. https://doi.org/10.1186/s12874-016-0165-8

Gossler, T., Sigala, I. F., Wakolbinger, T., & Buber, R. (2019). Applying the Delphi Method to Determine Best Practices for Outsourcing Logistics in Disaster Relief. *Journal of Humanitarian Logistics and Supply Chain Management*, *9*(3), 2042–6747. https://doi. org/10.1108/JHLSCM-06-2018-0044

Habibi, A., Sarafrazi, A., & Izadyar, S. (2015). Delphi Technique Theoretical Framework in Qualitative Research. *The International Journal Of Engineering And Science (IJES)* ||, *December*. www.theijes.com

Hallowell, M. R., & Gambatese, J. A. (2010). Qualitative Research: Application of the Delphi Method to CEM Research. *Journal of Construction Engi- Neering and Management*, *136*(1), 99–107. https://doi.org/10.1061/ASCECO.1943-7862.0000137

Hasson, F., Keeney, S., & McKenna, H. (2000). Research Guidelines for the Delphi Survey Technique. *Journal of Advanced Nursing*, *32*(4), 1008–1015. https://doi. org/10.1046/j.1365-2648.2000.t01-1-01567.x

Helmer, O., & Rescher, N. (1960). *On the Epistemology of the Inexact Sciences*. *Management Science*, 6(1), 25–52.

Holey, E. A., Feeley, J. L., Dixon, J., & Whittaker, V. J. (2007a). An Exploration of the Use of Simple Statistics to Measure Consensus and Stability in Delphi Studies. *BMC Medical Research Methodology*, *7*(February). https://doi.org/10.1186/1471-2288-7-52

Holey, E. A., Feeley, J. L., Dixon, J., & Whittaker, V. J. (2007b). An Exploration of the Use of Simple Statistics to Measure Consensus and Stability in Delphi Studies. *BMC Medical Research Methodology*, *7*(52). https://doi.org/10.1186/1471-2288-7-52

Jeon, Y. H., Conway, J., Chenoweth, L., Weise, J., Thomas, T. H., & Williams, A. (2015). Validation of a Clinical Leadership Qualities Framework for Managers in Aged Care: A Delphi Study. *Journal of Clinical Nursing*, *24*(7–8), 999–1010. https://doi.org/10.1111/ jocn.12682

Kattirtzi, M., & Winskel, M. (2020). When Experts Disagree: Using the Policy Delphi Method to Analyse Divergent Expert Expectations and Preferences on UK Energy Futures. *Technological Forecasting and Social Change*, *153*(February). https://doi.org/10.1016/j. techfore.2020.119924

Khoshfetrat, R., Sarvari, H., Chan, D. W. M., & Rakhshanifar, M. (2020). Critical Risk Factors for Implementing Building Information Modelling (BIM): A Delphi-Based Survey. *International Journal of Construction Management*. https://doi.org/10.1080/15623599.2 020.1788759

Kirschbaum, M., Barnett, T., & Cross, M. (2019). Q Sample Construction: A Novel Approach Incorporating a Delphi Technique to Explore Opinions about Codeine Dependence. *BMC Medical Research Methodology*, *19*(1), 1–13. https://doi.org/10.1186/s12874-019-0741-9

Kluge, U., Ringbeck, J., & Spinler, S. (2020). Door-to-Door Travel in 2035 – A Delphi Study. *Technological Forecasting and Social Change*, *157*, 120096. https://doi.org/10.1016/j. techfore.2020.120096

Lange, T., Kopkow, C., Lützner, J., Günther, K. P., Gravius, S., Scharf, H. P., Stöve, J., Wagner, R., & Schmitt, J. (2020). Comparison of Different Rating Scales for the Use in Delphi Studies: Different Scales Lead to Different Consensus and Show Different Test-Retest Reliability. *BMC Medical Research Methodology*, *20*(1), 1–12. https://doi.org/10.1186/ s12874-020-0912-8

Linstone, H. A., & Turoff, M. (2002). *The Delphi Method Techniques and Applications* (H. A. Linstone & M. Turoff (Eds.)).

Macphail, A. (2001). Nominal Group Technique: A Useful Method for Working with Young People. *Educational Research Journal*, *27*(2), 161–170. https://doi.org/10.1080/01411920120037117

McMillan, S. S., King, M., & Tully, M. P. (2016). How to Use the Nominal Group and Delphi Techniques. In *International Journal of Clinical Pharmacy* (Vol. 38, Issue 3, pp. 655–662). Springer Netherlands. https://doi.org/10.1007/s11096-016-0257-x

Milevska-kostova, N., & Dunn, W. N. (2010). Delphi Analysis. In Zaletel-Kragelj, L. & J. Bozikov (Eds.), *Methods and Tools in Public Health: A Handbook for Teachers, Researchers and Health Professionals* (pp. 423–436). Hans Jacobs Publishing Company.

Mokkink, L., Boers, M., Vleuten, C. van der, Bouter, L., Alonso, J., Patrick, D., Vet, H. de, & Terwee, C. (2020). COSMIN Risk of Bias Tool to Assess the Quality of Studies on Reliability or Measurement Error of Outcome Measurement Instruments: A Delphi Study. *BMC Medical Research Methodology*, *20*(293), 1–14. https://doi.org/10.21203/rs.3.rs-40864/v1

NSA. (2011). *The Delphi Technique* (Patent No. DOCID: 3928741). NSA.

Okoli, C., & Pawlowski, S. D. (2004). The Delphi Method as a Research Tool: An Example, Design Considerations and Applications. *Information and Management*, *42*(1), 15–29. https://doi.org/10.1016/j.im.2003.11.002

Schlecht, L., Schneider, S., & Buchwald, A. (2021). The Prospective Value Creation Potential of Blockchain in Business Models: A Delphi Study. *Technological Forecasting and Social Change*, *166*(July 2020). https://doi.org/10.1016/j.techfore.2021.120601

Schmalz, U., Spinler, S., & Ringbeck, J. (2021). Lessons Learned from a Two-Round Delphi-Based Scenario Study. *MethodsX*, *8*, 101179. https://doi.org/10.1016/j.mex.2020.101179

Skulmoski, G. J., Hartman, F. T., & Krahn, J. (2007). The Delphi Method for Graduate Research. *Journal of Information Technology Education*, *6*, 1–21.

Somiah, M. K. (2019). *An Integrated Competitive Advantage Model For Indigenous Construction Firms In The Ghanaian Construction Industry An Integrated Competitive Advantage Model For Indigenous Construction Firms* (Doctoral dissertation, University of Johannesburg (South Africa).

Tengan, C. (2018). *An Integrated Monitoring And Evaluation Model For Construction Project Delivery In Ghana.* (Issue July). University of Johannesburg.

Tetzlaff, J. M., Moher, D., & Chan, A. W. (2012). Developing a Guideline for Clinical Trial Protocol Content: Delphi Consensus Survey. *Trials*, *13*. https://doi.org/10.1186/1745-6215-13-176

Turoff, M. (1970). The Design of a Policy Delphi. *Technological Forecasting and Social Change*, *2*(2), 149–171. https://doi.org/10.1016/0040-1625(70)90161-7

Welty, G. (1973). Some Problems of Selecting Delphi Experts for Educational Planning and Forecasting Exercises. *California Journal of Educational Research*, *24*(3), 129–134.

Woodcock, T., Adeleke, Y., Goeschel, C., Pronovost, P., & Dixon-Woods, M. (2020). A Modified Delphi Study to Identify the Features of High Quality Measurement Plans for Healthcare Improvement Projects. *BMC Medical Research Methodology*, *8*(20). https://doi.org/10.1186/s12874-019-0886-6

Yousuf, M. I. (2007). Using Experts' Opinions Through Delphi Technique. *Practical Assessment, Research & Evaluation*, *12*(4), 1–8.

9 Conclusion and Recommendations

9.1 Introduction

This chapter is the last part of this book. It discusses the contribution and recommendation of the study based on the achieved study objectives. The developed building information modelling (BIM) maturity model for developing countries using the South African construction industry (SACI) as the case study provides a guide and roadmap for construction industry stakeholders. Also, it helps in identifying the important dimensions to be implemented to achieve optimum BIM maturity in developing countries. This is to ensure the full achievement of the BIM maturity attainment outcomes. Thus, the developed maturity model is critical to the technological diffusion in developing countries. It ensures transitioning from the slow and beginner phase of BIM in developing countries to ensure an accelerated BIM diffusion through a systematic approach to BIM maturity. It also ensures that there is a better, more efficient, and more productive construction process and a better product delivery to the clients. Furthermore, the developed model is holistic and thus impacts every aspect of the construction industry. Achieving BIM maturity is critical to the efficiency of the construction industry in the digital era, as most emerging technologies are dependent on data. Data and information management are the core of the BIM process; thus, it is critical to the construction industry transformation (Adekunle, Aigbavboa et al., 2021). It is worthy of note that this model is tailored to the developing countries context as maturity models are not a one size fits; they are context responsive. It is, therefore, critical to the BIM maturity of developing countries. This section discusses the contribution of this study and the recommendations

9.2 Contribution to Knowledge

The innovative value of the study is found in the development and validation of the BIM maturity model in the SACI. This study revealed that three variables are critical to achieving BIM maturity in the SACI context; these variables are effective BIM training, stakeholder orientation, and process. Other constructs considered and tested by the study from existing maturity models developed in different contexts are technology, ecosystem regulation, and people. These six constructs revealed the study's hypothesised BIM maturity model. The model revealed a holistic latent variable that has not been studied before now. Furthermore, the process

DOI: 10.1201/9781003373919-11

construct is rooted in the ISO 19650 standard. This is novel to BIM maturity model development, hence the developed model's uniqueness.

Evidence from the literature reveals that BIM maturity has been subjected to different shortcomings, as earlier stated; this includes non-documentation of the model development process and similar methodologies of either an interview or a quantitative data collection approach. The Delphi survey adopted by the study achieved an articulated exploration of the BIM maturity model in the SACI. Also, the study modelled a holistic six construct that consists of three existing constructs from literature, namely, process, people, and technology, and three new gaps, namely, ecosystem regulation, effective BIM training, and stakeholder orientation. These are new methodological addition to the BIM maturity model development. Consequently, this affords other similar studies to adopt.

Most of the previous studies on BIM maturity model development failed to provide any evidence of the model development and validation in the literature. This study presents a documented and well-articulated methodological approach. At each stage of analysis, the constructs and the observed variables were analysed to determine their validity and reliability. The Delphi survey analysis was conducted to define consensus, which provided validity and reliability through the iterative nature of the process.

9.3   Industry Contributions and Values

The findings of this study are very useful to the industry practitioner and stakeholders towards achieving BIM maturity. Firstly, the study presents a practical and verifiable methodology to achieve its aims. The adoption of Delphi ensured that an in-depth determination of the factors for BIM maturity was achieved. The Delphi experts for the study were heterogeneous.

Secondly, the present challenges of achieving BIM maturity in SACI were provided. These study findings identified the critical factors for BIM maturity in the SACI: process, effective BIM training, and stakeholder orientation. Hence, it provides the areas that require prioritisation by industry stakeholders to achieve BIM maturity.

Thirdly, this study provides a BIM maturity architecture that the SACI can adopt to achieve maturity. This architecture is systematic and provides a framework for BIM maturity in the SACI.

Finally, this study can also be adopted by other developing countries because the SACI adopted for this study has some similarities with other developing countries. It is presently experiencing low BIM maturity like other developing countries (Adekunle, Ejohwomu, & Aigbavboa, 2021). Also, it lacks a government mandate towards BIM adoption. Furthermore, its economic climate might not be the same as other developing countries; however, it is not well developed like the developed countries. Hence, the study is useful for other developing countries.

9.4   Recommendations

From the preceding, the study makes the following recommendations from theoretical, methodological, and industry perspectives.

9.4.1 Recommendations

Literature on BIM maturity model development has been observed to have similar approaches (Kassem et al., 2020), hence the similar results, methodological underpinnings, and absence of documentation. This study adopted systematic and novel approaches in the research area. This study has six constructs for BIM maturity development, which are not similar to those adopted in previous studies. It is therefore recommended that similar studies should adopt scientific and systematic approaches to developing BIM maturities.

It is recommended that maturity model development studies in other digitisation fields for new technologies adopt a similar methodological approach. It should also be adopted in the maturity model development research area as it enables a better exploration and robustness of the data collected.

9.4.2 Industry Recommendation

The following recommendations are for industry stakeholders to ensure achieving BIM maturity:

1 Industry stakeholders should focus on process enhancement to be BIM tailored, proper orientation of the stakeholders on BIM benefits, and effective BIM training. This will achieve a faster BIM maturity.
2 Working in silos will not achieve BIM maturity in SACI, but a collaborative approach will accomplish this.
3 Adopting SWOT thinking, stakeholders should focus on the three critical factors to ensure BIM maturity faster than other factors presently being focused on.
4 The SACI should adopt the BIM maturity model architecture proposed by the study for a systematic BIM maturity of the SACI.

As earlier stated, other developing countries should consider adopting the developed BIM maturity model in their contexts to achieve BIM maturity.

References

Adekunle, S. A., Aigbavboa, C. O., Ejohwomu, O., Adekunle, E. A., & Thwala, W. D. (2021). Digital Transformation in the Construction Industry : A Bibliometric Review. *Journal of Engineering, Design and Technology*, *2013*. https://doi.org/10.1108/JEDT-08-2021-0442
Adekunle, S. A., Ejohwomu, O., & Aigbavboa, C. O. (2021). Building Information Modelling Diffusion Research in Developing Countries: A User Meta-Model Approach. *Buildings*, *11*(7), 264. https://doi.org/10.3390/buildings11070264
Kassem, M., Li, J., Kumar, B., Malleson, A., Gibbs, D. J., Kelly, G., & Watson, R. (2020). *Building Information Modelling: Evaluating Tools for Maturity and Benefits Measurement*. Centre for Digital Built Britain, 184.

Index

Note: **Bold** page numbers refer to tables and *italic* page numbers refer to figures.

For Product Safety Concerns and Information please contact our EU
representative GPSR@taylorandfrancis.com
Taylor & Francis Verlag GmbH, Kaufingerstraße 24, 80331 München, Germany

www.ingramcontent.com/pod-product-compliance
Lightning Source LLC
Chambersburg PA
CBHW060317220326
41598CB00027B/4353

* 9 7 8 1 0 3 2 4 4 7 8 9 6 *